NONCIRCULAR GEARS

Noncircular Gears: Design and Generation represents the extension of the modern theory of gearing applied to the design and manufacture of the main types of noncircular gears: conventional and modified elliptical gears, eccentric gears, oval gears, gears with lobes, and twisted gears. This book is enhanced by updated theoretical descriptions of the methods of generation of noncircular gears by enveloping methods similar to those applied to the generation of circular gears. *Noncircular Gears: Design and Generation* also offers new developments intended to extend the application of noncircular gears for output speed variation and generation of functions. Numerous numerical examples show the application of the developed theory. This book aims to extend the application of noncircular gear drives in mechanisms and industry.

Faydor L. Litvin has been a Professor at the University of Illinois at Chicago for the past 30 years, after 30 years as a Professor and Department Head of Leningrad Polytechnic University and Leningrad Institute of Precise Mechanics and Optics. Dr. Litvin is the author of more than 300 publications (including 10 monographs) as well as the inventor and co-inventor of 25 inventions. Among his many honors, Dr. Litvin was made Doctor Honoris Causa of Miskolc University, Hungary, in 1999. He was named Inventor of the Year 2001 by the University of Illinois at Chicago and has been awarded 12 NASA Tech Brief awards; the 2001 Thomas Bernard Hall Prize (Institution of Mechanical Engineers, UK); and the 2004 Thomas A. Edison Award (ASME). He was elected a Fellow of ASME and is an American Gear Manufacturers Association (AGMA) member. He has supervised 84 Ph.D. students. In addition to his deep interest in teaching, Dr. Litvin has conducted seminal research on the theory of mechanisms and the theory and design of gears.

Alfonso Fuentes-Aznar is a Professor of Mechanical Engineering at the Polytechnic University of Cartagena (UPCT). He has more than 15 years of teaching experience in machine element design and is the author of more than 65 publications. He was a Visiting Scholar and Research Scientist at the Gear Research Center of the University of Illinois at Chicago from 1999 to 2001. Among his many honors, he was awarded the 2001 Thomas Bernard Hall Prize (Institution of Mechanical Engineers, UK) and the NASA Tech Brief Award No. 17596-1 for the development of a new technology – "New Geometry of Face Worms Gear Drives with Conical and Cylindrical Worms." He is a member of the editorial boards of *Mechanism and Machine Theory*, the *Open Mechanical Engineering Journal*, and *Recent Patents on Mechanical Engineering*. Dr. Fuentes-Aznar is a member of AGMA and ASME. This is his second book.

Ignacio Gonzalez-Perez earned his Ph.D. from the Polytechnic University of Cartagena. He was a Visiting Scholar at the Gear Research Center of the University of Illinois at Chicago, from 2001 to 2003 and a Visiting Scientist at the Gear Research and Development Department of Yamaha Motor Co., Ltd., in Japan, in 2005. Dr. Gonzalez-Perez is currently an Associate Professor in the Department of Mechanical Engineering of the Polytechnic University of Cartagena. He is a member of AGMA and ASME. He received an AGMA student paper award in 2002 and a UPCT research award for his thesis in 2006.

Kenichi Hayasaka graduated from Tohoku University, Japan, in 1977 and earned his master's degree in 1983 from the University of Illinois at Chicago. He is the author of 16 research publications and inventor and co-inventor of 23 inventions. Mr. Hayasaka has been a researcher and engineer at Yamaha Motor Co., Ltd., since 1977. He is involved in the research and design of gears and transmissions applied in motorcycles and marine propulsion systems, and in the development of enhanced computerized design systems for optimizing low noise and durability of spiral bevel gears, helical gears, and spur gears.

Noncircular Gears

DESIGN AND GENERATION

Faydor L. Litvin
University of Illinois at Chicago

Alfonso Fuentes-Aznar
Polytechnic University of Cartagena

Ignacio Gonzalez-Perez
Polytechnic University of Cartagena

Kenichi Hayasaka
Yamaha Motor Co., Ltd.

CAMBRIDGE UNIVERSITY PRESS
Cambridge, New York, Melbourne, Madrid, Cape Town, Singapore,
São Paulo, Delhi, Dubai, Tokyo

Cambridge University Press
32 Avenue of the Americas, New York, NY 10013-2473, USA

www.cambridge.org
Information on this title: www.cambridge.org/9780521761703

First published 2009

Printed in the United States of America

A catalog record for this publication is available from the British Library.

Library of Congress Cataloging in Publication Data

Noncircular gears : design and generation / Faydor L. Litvin ... [et al.].
     p.   cm.
Includes bibliographical references and index.
ISBN 978-0-521-76170-3 (hardback)
1. Gearing.   2. Ellipsoid.   I. Litvin, F.L. (Faydor L.)   II. Title.
TJ184.N66   2009
621.8′33 – dc22   2009011763

ISBN 978-0-521-76170-3 Hardback

To my followers:

When I'll be very far,
And look at you from a twinkling star,
I'll whisper (will you hear?)
I love you, my very dear.

FAYDOR L. LITVIN

Contents

Foreword   *page* xi

Preface   xiii

Acknowledgments   xv

1 **Introduction to Theory of Gearing, Design, and Generation of
  Noncircular Gears** 1

   1.1 Historical Comments   1
   1.2 Toward Design and Application of Noncircular Gears   6
      1.2.1 Examples of Previous Designs   6
      1.2.2 Examples of New Designs   11
   1.3 Developments Related with Theory of Gearing   14

2 **Centrodes of Noncircular Gears** 18

   2.1 Introduction   18
   2.2 Centrode as the Trajectory of the Instantaneous Center of
      Rotation   20
   2.3 Concept of Polar Curve   21
   2.4 Derivation of Centrodes   21
   2.5 Tangent to Polar Curve   22
   2.6 Conditions for Design of Centrodes as Closed Form Curves 24
   2.7 Observation of Closed Centrodes for Function Generation   27
   2.8 Basic and Alternative Equations of Curvature of Polar Curve 27
   2.9 Conditions of Centrode Convexity   30

3 **Evolutes and Involutes** 31

   3.1 Introduction and Terminology   31
   3.2 Determination of Evolutes   33
   3.3 Local Representation of a Noncircular Gear   36
   3.4 Pressure Angle   37

4 **Elliptical Gears and Gear Drives** 40
   4.1 Introduction   40
   4.2 Basic Concepts   40
        4.2.1 Ellipse Parameters   40
        4.2.2 Polar Equation of an Ellipse   41
   4.3 External Elliptical Gear Drives   43
        4.3.1 Basic Equations   43
        4.3.2 Conventional Elliptical Gear Drives   45
               4.3.2.1 Centrodes and Transmission Function   45
               4.3.2.2 Influence of Ellipse Parameters and Design
                        Recommendations   47
        4.3.3 Gear Drive with Elliptical Pinion and Conjugated Gear   52
        4.3.4 External Gear Drive with Modified Elliptical Gears   53
               4.3.4.1 Modification of the Ellipse   53
               4.3.4.2 Derivation of Modified Centrode σ_1 55
               4.3.4.3 Derivative Functions $m_{21}^{(I)}(\phi_1)$ and $m_{21}^{(II)}(\phi_1)$ 55
               4.3.4.4 Relation between Rotations of Gears 1 and 2   55
               4.3.4.5 Derivation of Centrode σ_2 56
        4.3.5 External Gear Drive with Oval Centrodes   57
               4.3.5.1 Equation of Oval Centrode   57
               4.3.5.2 Derivative Function $m_{21}(\phi_1)$ 58
               4.3.5.3 Relation between Rotations of Gears 1 and 2   58
               4.3.5.4 Transmission Function $\phi_2(\phi_1)$ 60
        4.3.6 Design of Noncircular Gears with Lobes   60
               4.3.6.1 Design of Gear Drives with Different Number of
                        Lobes for Pinion and Gear   63
   4.4 Transmission Function of Elliptical Gears as Curve of Second
        Order   65
   4.5 Functional of Identical Centrodes   66

5 **Generation of Planar and Helical Elliptical Gears** 71
   5.1 Introduction   71
   5.2 Generation of Elliptical Gears by Rack Cutter   71
   5.3 Generation of Elliptical Gears by Hob   79
   5.4 Generation of Elliptical Gears by Shaper   86
   5.5 Examples of Design of Planar and Helical Elliptical Gears   90
        5.5.1 Planar Elliptical Gears   90
        5.5.2 Helical Elliptical Gears   92

6 **Design of Gear Drives Formed by Eccentric Circular Gear and
   Conjugated Noncircular Gear** 94
   6.1 Introduction   94
   6.2 Centrodes of Eccentric Gear Drive   94

6.2.1   Equations of Mating Centrodes   94
6.2.2   Curvature of Centrode σ_2 and Applications   96
6.3   Generation of the Noncircular Gear by Shaper and Hob   101
6.3.1   Generation of Noncircular Gear by a Noneccentric Shaper   101
6.3.2   Generation of the Noncircular Gear by a Hob   105
6.4   Generation of the Eccentric Gear Providing Localized Contact   112

7   **Design of Internal Noncircular Gears** 115
7.1   Introduction   115
7.2   Derivation of Centrodes   115
7.2.1   Preliminary Considerations of Kinematics of Internal Gear
        Drive   115
7.2.2   Basic Equations of Centrodes   116
7.2.3   Design of Centrodes σ_1 and σ_2 as Closed-Form Curves   118
7.3   Examples of Design of Internal Noncircular Gear Drives   118
7.3.1   Gear Drive with Elliptical Pinion   118
7.3.2   Gear Drive with Modified Elliptical Pinion   120
7.3.3   Gear Drive with Oval Pinion   121
7.3.4   Gear Drive with Eccentric Pinion   123
7.4   Generation of Planar Internal Noncircular Gears by Shaper   126
7.5   Conditions of Nonundercutting of Planar Internal Noncircular
      Gears   132
7.5.1   Approach A   133
7.5.2   Approach B   134
7.5.3   Numerical Example   136

8   **Application for Design of Planetary Gear Train with Noncircular
    and Circular Gears** 138
8.1   Introduction   138
8.2   Structure and Basic Kinematic Concept of Planetary Train   138
8.3   Planetary Gear Train with Elliptical Gears   139
8.4   Planetary Gear Train with Noncircular and Circular Gears   141

9   **Transformation of Rotation into Translation with Variation of
    Gear Ratio** 143
9.1   Introduction   143
9.2   Determination of Centrodes of Noncircular Gear and Rack   144
9.3   Application of Mechanism Formed by a Noncircular Gear and
      Rack   144

10 **Tandem Design of Mechanisms for Function Generation and Output
    Speed Variation** 147
10.1 Introduction   147
10.2 General Aspects of Generation of Functions   152

10.3 Generation of Function with Varied Sign of Derivative   153
10.4 Introduction to Design of Multigear Drive with Noncircular
     Gears   156
     10.4.1 Basic Functionals   156
     10.4.2 Interpretations of Lagrange's Theorem   160
     10.4.3 Illustration of Application of Lagrange's Theorem for
            Functional $\psi(\alpha) = g_2(g_1(\alpha))$ 161
            10.4.3.1 Previous Solutions for $\psi(\alpha) = f(f(\alpha))$ 161
            10.4.3.2 Computational Procedure for Functional (10.4.6)   163
10.5 Design of Multigear Drive   166
     10.5.1 Basic Equations   166
     10.5.2 Design of Centrodes   168
10.6 Design of Planar Linkages Coupled with Noncircular Gears   171
     10.6.1 Tandem Design of Double-Crank Mechanism Coupled
            with Two Pairs of Noncircular Gears   171
     10.6.2 Tandem Design of Slider-Crank Mechanism Coupled with
            Modified Elliptical Gears   174
            10.6.2.1 Preliminary Information   174
            10.6.2.2 Basic Ideas of Modification of Elliptical Centrodes   174
            10.6.2.3 Analytical Determination of Modified Elliptical
                      Centrodes   177
            10.6.2.4 Numerical Examples   179
     10.6.3 Tandem Design of Scotch-Yoke Mechanism Coupled with
            Noncircular Gears   180
     10.6.4 Tandem Design of Mechanism Formed by Two Pairs of
            Noncircular Gears and Racks   183
            10.6.4.1 Generation of Function   186

11 **Additional Numerical Problems** 188

References   201
Index   203

Foreword

Noncircular gears (NCG) have been considered a curiosity and a product of niche applications for a long time because of their design and manufacture complexity. However, thanks to the availability of powerful computers and sophisticated CNC machine tools, the design and manufacturing of NCG became more feasible and their potential in many fields can be more easily exploited.

NCG are used to improve the function, versatility, and simplicity of many mechanical processes. For instance, they allow speed matching on assembly lines, linear motion with quick return, and stop-and-dwell motion. Very recent publications in highly respected scientific journals show that the interest in NCG is still very strong, for both the theoretical and manufacturing challenges involved, especially in today's high-performance automatic machines.

In this context, this book on NCG is very welcome. It covers all the basic issues of NCG and shows how to solve most of the key problems in NCG design and manufacturing. It clearly presents the foundations of this topic, very useful to the beginner, as well as new and advanced concepts of great importance to both scientists involved in research issues and skillful designers facing NCG sizing and manufacturing. In particular, the book represents the most comprehensive and authoritative treatise on NCG available today for scientists and designers. When complemented with a recently published book, *Gear Geometry and Applied Theory* (2nd ed.) by F. L. Litvin and A. Fuentes (Cambridge University Press, 2004), which provides the fundamentals of all types of gears, the whole field of the theory and applications of gears is covered.

This book on NCG is a bridge between the past and the future of this advanced topic, representing the extension of the existing theory of gearing as it has been developed by Prof. Litvin and his co-authors. In particular, the main contributions of the book can be summarized as follows:

(a) A new algorithm is presented for generation of a given function by any number of gear pairs. It allows more favorable shapes of gear centrodes to be obtained. The analytical solution of this topic did not exist previously.

(b) Design of many types of noncircular gear drives comprising conventional and modified elliptical gears, eccentric circular gears, oval gears, gears with lobes, or internal noncircular gears is covered and their methods of generation by rack cutter, hob, or shaper presented.

(c) New approaches for the analytical design of centrodes are presented, such as modified elliptical centrodes applied for external and internal gear drives. Most of the transmission functions are represented analytically, allowing conditions for avoiding undercutting and interference to be obtained.

(d) Finally, tandem design as a combination of a linkage and a gear drive with NCG, as represented in the book, extends the possibility of variation of the output velocity.

Of Prof. Litvin's life and scientific activity much has been said. In particular, after the most striking sentence reported by John J. Coy (*Mechanism and Machine Theory*, Vol. 30, n. 3, pp. 491–492, 1995) in celebration of the eighty-year milestone of Prof. Litvin, which enlightens his extraordinary personality, not too much remains to be added. Indeed, Dr. Coy reported: "At the time when most people are content to put up their feet and rest, Dr. Faydor Litvin has other things he would rather do. ... He is always quick to thank others for their inspiration and support. He is grateful for the health and vigor that has enabled his continued contribution." Now, at the age of 95, he and his co-authors present this new and needed book.

I am not the first to believe that Prof. Litvin is an exceptional scientist and a talented man. He faced the writing of this book with the enthusiasm of a young researcher preparing his first paper, but with the wisdom and high depth of this respectable age spent studying and advancing the modern theory of gears.

The reader will be happy and soon fascinated by this book, and will certainly owe Prof. Litvin and his co-authors for such a remarkable piece of work.

Prof. Vicenzo Parenti-Castelli
University of Bologna, Italy

Preface

The book is written by a group of authors united by application of the same methodology and experience of cooperation for a long time. The contents of the book cover:

(1) Methods of generation of noncircular gears by enveloping methods that are similar to those applied for generation of circular gears. However, the motions of the generating tools (rack cutter, shaper, or hob) are nonlinear and must be computerized.

(2) Design of noncircular gears to be applied for variation of output speed and generation of functions. Such a design requires determination of mating centrodes that roll over each other.

(3) Detailed procedure of design of elliptical gears with spur and helical teeth, oval gears, lobes, eccentric involute gears, and twisted gears (applied for extension of the interval of function generation).

(4) Tandem design of planar linkages coupled with noncircular gears for a broader variation range of the output speed.

The authors hope that this book will allow extension of application and design of noncircular gear drives in mechanisms and industry.

Faydor L. Litvin
Alfonso Fuentes-Aznar
Ignacio Gonzalez-Perez
Kenichi Hayasaka

Acknowledgments

The authors express their deep gratitude to institutions, companies, and colleagues that have supported their research. Special thanks to Mr. Kenji Yukishima and Mr. Hiroyuki Nagamoto for their valuable contributions to the preparation of this book.

1 Introduction to Theory of Gearing, Design, and Generation of Noncircular Gears

1.1 Historical Comments

Designers have tried for many years to apply noncircular gears in automatic machines and instruments. The obstacle was the lack of effective methods of generation of noncircular gears similar to those applied for the generation of circular gears. However, researchers have continued the investigation of application of noncircular gears – see the earlier works by Burmester (Burmester, 1888), Golber (Golber, 1939), Temperley (Temperley, 1948), or Boyd (Boyd, 1940) – and manufacturers have intensified their efforts for improvement of the generation of noncircular gears (Fellows, 1924; Bopp & Reuther G.m.b.H., 1938).

Due to the lack of exact methods of generation of noncircular gears, the efforts were first directed to the development of methods based on the meshing of generating tools with master gears. Figure 1.1.1 shows the Fellows' approach where the noncircular master gear 1 is in mesh with a master rack (Fellows, 1924). The rack cutter and gear being generated are denoted by 3 and 4, respectively.

Bopp and Reuther's approach (Fig. 1.1.2) is based on the simulation of meshing of a noncircular master worm gear c with a worm f that is identical to the hob d; a is the spur noncircular gear being generated; the cam b and the follower e form the cam mechanism designated for simulation of the required variable distance between c and f (Bopp & Reuther G.m.b.H., 1938). Weight g maintains the continuous contact between the cam and the follower. However, both approaches were difficult to apply due to the necessity of manufacturing noncircular master gears, which was expensive and time-consuming.

The breakthrough of generation of noncircular gears happened in years 1949–1951, wherein enveloping methods of generation of noncircular gears were developed based on generation by a rack cutter, hob, or a shaper (Litvin *et al.*, 1949 to 1951). Such methods were based on obtaining the gear tooth surface as the envelope to the family of tool surfaces and are based on the following ideas:

(a) The noncircular gears are generated by the same tools (rack cutters, hobs, and shapers) that are used for the manufacture of circular gears.

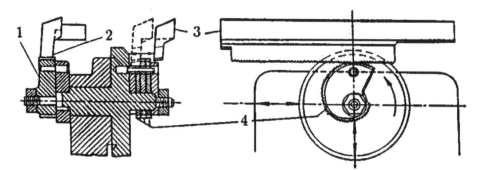

Figure 1.1.1. Generation of a noncircular gear by application of (i) a master rack 2; (ii) a master noncircular gear 1; (iii) a cutting rack cutter 3. The noncircular gear being generated is 4.

(b) Conjugated tooth shapes for noncircular gears are provided due to the imaginary rolling of the tool centrode over the given gear centrode.

(c) The imaginary rolling of the tool centrode over the centrode of the gear being generated is accomplished by proper relations between the motions of the tool and the gear in the process of cutting.

We illustrate the developed approaches in Fig. 1.1.3, which shows that mating noncircular centrodes 1 and 2 are in mesh with a conventional rack cutter 3. The centrode of the rack cutter is a straight line $\overline{t-t}$ that is a common tangent to centrodes 1 and 2 and rolls over 1 and 2. Rolling of centrode 3 of the rack cutter is

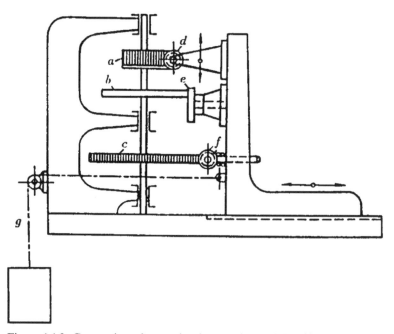

Figure 1.1.2. Generation of a noncircular gear by applying (i) a master worm gear c being in mesh with worm f; (ii) a cam b and follower e; (iii) a is the gear being generated; (iv) d is the hob.

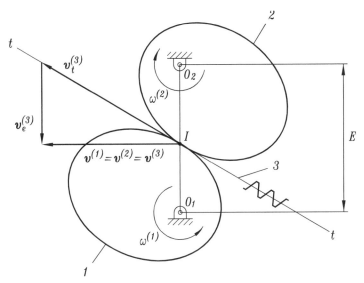

Figure 1.1.3. Illustration of generation of noncircular gears 1 and 2 by rack cutter 3.

achieved wherein the rack cutter translates along tangent $\overline{t-t}$ and is rotated about the instantaneous center of rotation I (Fig. 1.1.3). Tooth surfaces of gear 1, gear 2, and rack cutter 3 are in mesh simultaneously, and gear 1 and 2 are provided with conjugated surfaces.

Drawings of Fig. 1.1.4 show the related motions of a rack cutter being in mesh with one of the noncircular gears of the pair of noncircular gears shown in Fig. 1.1.3. Figure 1.1.4(a) shows that the rack cutter translates along $\overline{t-t}$, which is the common tangent to the rack cutter centrode 2 and centrode 1 of the noncircular gear. Centrode 2 is a straight line. Centrode 1 applied for generation is the same as that applied in the meshing of two mating noncircular gears, as shown in Fig. 1.1.3.

The noncircular gear 1 being in mesh with rack cutter 2 performs two related motions during the process of generation (Fig. 1.1.4(a)): (a) rotation about O_1 with angular velocity $\omega^{(1)}$, and (b) translational motion with linear velocity $\mathbf{v}_{tr}^{(1)}$ in a direction that is perpendicular to $\overline{t-t}$. Rolling of the rack cutter 2 about centrode 1 is provided by observation of the vector equation

$$\mathbf{v}^{(2)} = \mathbf{v}_{rot}^{(1)} + \mathbf{v}_{tr}^{(1)} \tag{1.1.1}$$

We may consider that the translation of the rack cutter is performed with a constant velocity $\mathbf{v}^{(2)}$, and the motions of the noncircular gear are provided by observation of the nonlinear function $\boldsymbol{\omega}^{(1)}(\mathbf{v}^{(2)})$ and $\mathbf{v}_{tr}^{(1)}(\mathbf{v}^{(2)})$.

Three coordinate systems are applied for generation: (i) movable ones, S_2 and S_1, rigidly connected to the rack cutter 2 and noncircular gear 1 (Fig. 1.1.4(b)), and (ii) fixed coordinate system S_f, in which we consider the motions of S_2 and S_1 (Fig. 1.1.4(b)). Symbols $x_f^{(O_2)}$ and $y_f^{(O_1)}$ denote the displacements of the rack cutter and noncircular gear, respectively. Angle ϕ_1 denotes the rotation of the gear.

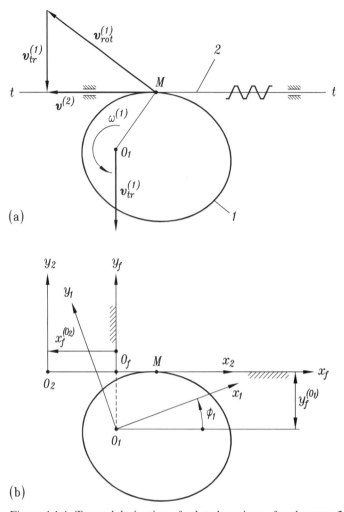

Figure 1.1.4. Toward derivation of related motions of rack cutter 2 and noncircular gear 1.

The derivations of the nonlinear functions $\phi_1(x_f^{(O_2)})$ and $y_f^{(O_1)}(x_f^{(O_2)})$ are presented in Chapter 5. Initially, observation of functions mentioned previously has been accomplished by the remodeling of existing equipment and using cam mechanisms for the generation of the required functions. Figure 1.1.5 shows the first cutting machine for noncircular gears applied in 1951. At present, observation of functions $\phi_1(x_f^{(O_2)})$ and $y_f^{(O_1)}(x_f^{(O_2)})$ is obtained by computer controlled machines (Smith, 1995).

By using enveloping methods of generation of noncircular gears, various new types of noncircular gears have been developed with closed and non-closed centrodes. An example of a pair of noncircular gears with non-closed centrodes applied for generation of a given function $y(x)$ represented in the closed interval $[x_1, x_2]$, if $y(x_1) \neq y(x_2)$ (see Chapter 10), is shown in Fig. 1.1.6.

Figure 1.1.7 shows a 3D model of a helical elliptical gear. It has found good application for the driving of a crank-slider mechanism in heavy-press machines.

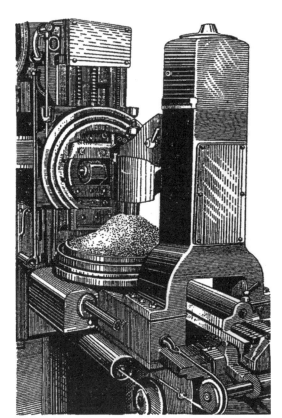

Figure 1.1.5. Remodeled cutting machine for generation of noncircular gears (1951).

Figure 1.1.6. Illustration of noncircular gear drive with non-closed centrodes.

Figure 1.1.7. Illustration of helical elliptical gear drive.

Figure 1.1.8 shows a gear drive formed by an eccentric involute pinion and a conjugated noncircular gear. Application of this gear drive has been found in rice planting machines. It can be used as well in tandem design of the Scotch-Yoke mechanism coupled with eccentric involute gears to obtain an improved function of the output velocity.

1.2 Toward Design and Application of Noncircular Gears

1.2.1 Examples of Previous Designs

Noncircular gears have found application in the industry for (i) variation of the output speed (for instance, in presses, conveyers, rice planting machines, etc.) and

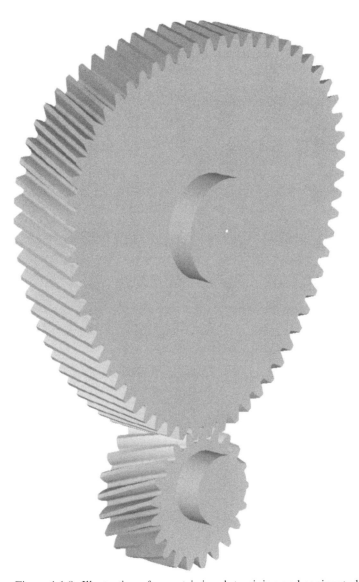

Figure 1.1.8. Illustration of eccentric involute pinion and conjugated noncircular gear.

(ii) generation of a given function (by a single pair of centrodes or a multigear drive).

The contents of this section cover only a small number of examples of the previous design of mechanisms with noncircular gears. Figure 1.2.1 shows application of a gear drive with elliptical gears applied for driving a Maltese cross mechanism. The Maltese cross mechanism, also known as the Geneva mechanism, is used to convert a continuous rotary motion into an intermittent rotary motion, and it is applied in many instruments or in other applications where an intermittent rotary motion is required. The purpose of the design shown in Fig. 1.2.1 is obtaining lower and higher speeds for the working and free-running parts of the cycle.

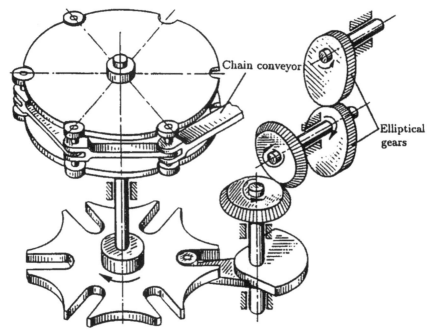

Figure 1.2.1. Application of elliptical gears in combination with a Maltese cross mechanism.

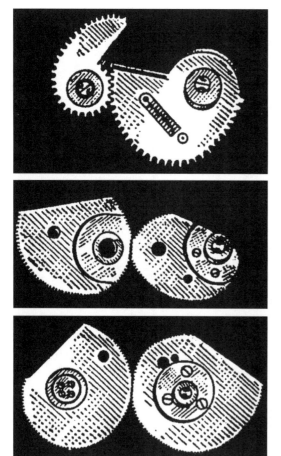

Figure 1.2.2. Illustration of noncircular gears applied in fine mechanics.

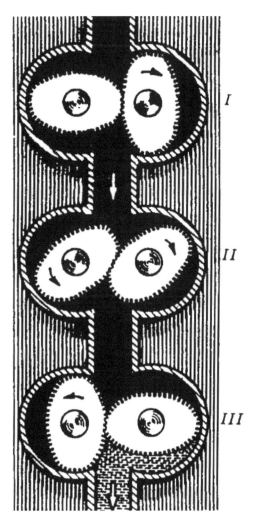

Figure 1.2.3. Illustration of transportation of the flow of liquid: I, II, and III positions of the same pair of oval gears.

Figure 1.2.2 shows examples of noncircular gears applied in the past in fine mechanics for the generation of functions of one variable. The development of electronic ways of generation of functions has eliminated such an approach. Figure 1.2.3 shows a comparatively modern example of the application of oval gears in flowmeters, proposed by Bopp and Reuthers (Bopp & Reuther G.m.b.H., 1938). The oval shape of the centrodes is obtained by modification of a conventional ellipse (see Section 4.3.5).

Figure 1.2.4 illustrates the possibility of application of twisted centrodes, which allow an increasing of the interval of the function to be generated. The gears with such centrodes may perform during the process of meshing an angle of rotation $\phi > 2\pi$ but, in addition to rotation, the gears have to perform an axial displacement. Figure 1.2.5 shows the centrodes of gears for generation of function $y(x) = 1/x$ for $1 \le x \le 3$ and $\phi_{1_{max}} = 5\pi$.

Figure 1.2.6 is the sketch of a heavy press machine designed as a combination of a crank-slider linkage coupled with elliptical gears. Such a design provides two

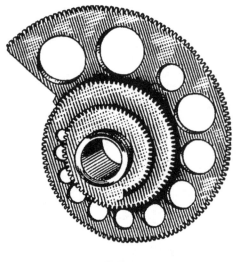

Figure 1.2.4. Illustration of a twisted noncircular gear.

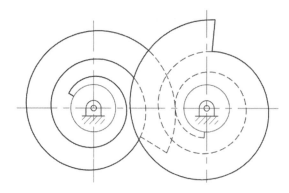

Figure 1.2.5. Centrodes of gears for generation of the function $y(x) = 1/x$ for $1 \leq x \leq 3$ and $\phi_{1max} = 5\pi$.

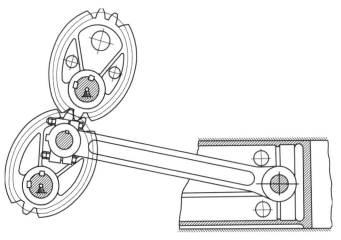

Figure 1.2.6. Application of elliptical gears in combination with a crank-linkage mechanism for a heavy press machine.

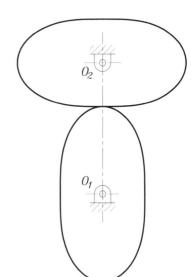

Figure 1.2.7. Illustration of oval gears.

parts of the cycle of the slider, with a lower and higher speed (for the working and free-running parts of the cycle). For applications in which there is a need to obtain a nonsymmetrical cycle of the slider, with a slow-working cycle and a fast free-running cycle, application of modified elliptical gears will be considered.

1.2.2 Examples of New Designs

New developments based on the modification of conventional ellipses have been accomplished. These new designs cover oval gears (Fig. 1.2.7), gears with lobes[1] (Fig. 1.2.8), or conjugation of an ellipse with a driven gear wherein the driving gear 1 performs more than one revolution for each revolution of the driven gear (Fig. 1.2.9).

The following examples illustrate the advantages of the tandem design of mechanisms with noncircular gears. Figure 1.2.10 shows the tandem design of a slider-crank linkage and a pair of modified elliptical gears to be applied in those applications wherein differences in the governing function corresponding to the approach stroke and to the return stroke is the goal. Figure 1.2.10(a) shows the centrodes of the modified elliptical gears, and Fig. 1.2.10(b) shows the unmodified $s(\beta)$ and modified $s(\alpha)$ governing functions of the slider-crank linkage. In this case, a longer approach stroke for the slider is obtained with a fast return stroke of the slider.

Another example of a new design is the Scotch-Yoke mechanism (Fig. 1.2.11(a)) coupled with noncircular gears for the variation of output speed. Figure 1.2.11(b) shows the relation between velocities, and Fig. 1.2.12 shows the shape of a transported textile ball as a function of $v_7(\phi_1)$.

---

[1] Lobe is a synonym of *part*.

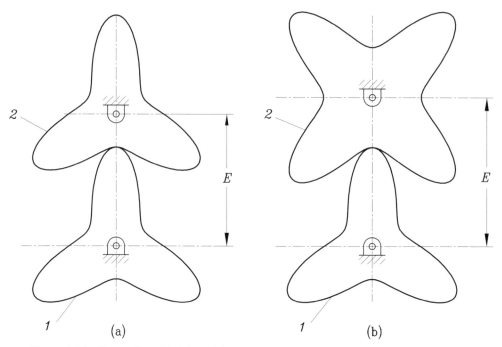

Figure 1.2.8. Centrodes with lobes: (a) three lobes and gear ratio equal to 1; (b) three and four lobes of centrodes 1 and 2, and gear ratio equal to 4/3.

A significant achievement is the generation of a given function by a multigear drive. The theoretical basis of this approach is the analytical solution of the functional

$$\psi(\alpha) = g_n(g_{n-1}(g_{n-2}(\ldots g_1(\alpha)))) \tag{1.2.1}$$

which is proposed in Chapter 10.

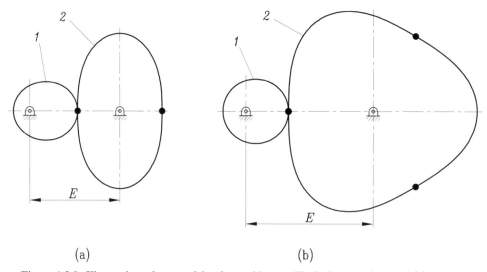

Figure 1.2.9. Illustration of a gear drive formed by an elliptical centrode 1 and (a) centrode 2 with two lobes; (b) centrode 2 with three lobes.

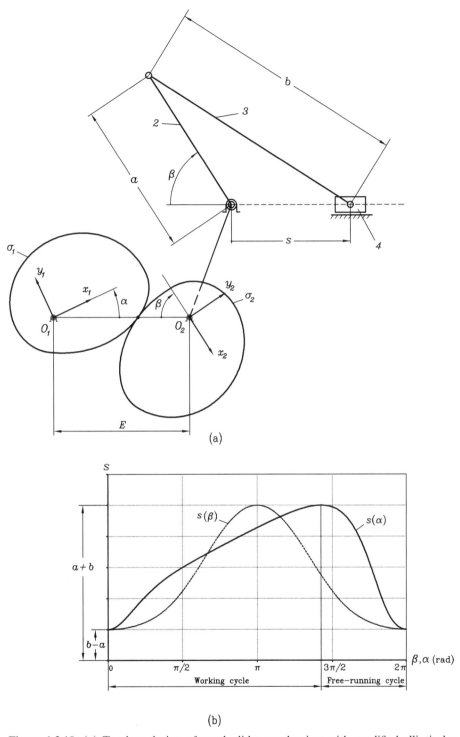

Figure 1.2.10. (a) Tandem design of crank-slider mechanism with modified elliptical cen-trodes; (b) unmodified $s(\beta)$ and modified $s(\alpha)$ governing functions of the geared crank-slider mechanism.

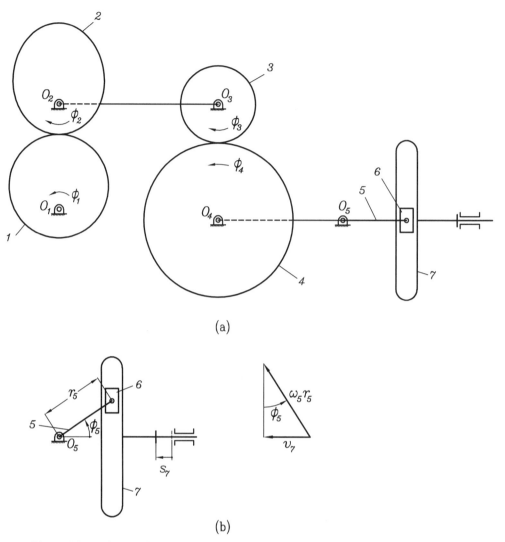

(a)

(b)

Figure 1.2.11. Composition of gear drive with Scotch-Yoke mechanism: (a) eccentric gear drive and Scotch-Yoke mechanism at their initial positions; (b) links of Scotch-Yoke mechanism and details of related velocities.

Figure 1.2.13 shows the case when the given function $\psi(\alpha)$ is generated by three pairs of unclosed centrodes. The figure illustrates a particular case of solution of the functional (1.2.1). The great advantage of application of a multigear drive formed by noncircular gears is the reduction of the pressure angle of the centrodes and their favorable shape (see Chapter 10).

1.3 Developments Related with Theory of Gearing

Teeth of the same noncircular gear have different profiles, and this is the main difference between a circular and noncircular gear. The geometry of teeth and their profiles must be based on the concepts applied in differential geometry for planar curves.

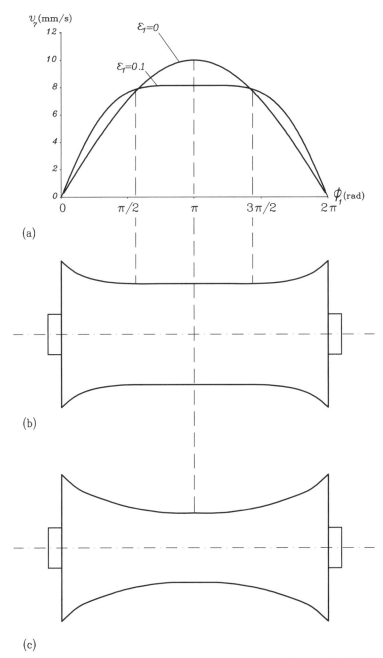

Figure 1.2.12. Tandem design of Scotch-Yoke mechanism coupled with eccentric gear drive: (a) velocity $v_7(\phi_1)$ as function of eccentricity ε_1; (b) shape of textile ball corresponding to $\varepsilon_1 = 0.1$; (c) shape of textile ball wherein noncircular gears are not applied ($\varepsilon_1 = 0$).

TANGENT TO CENTRODES. The consideration of centrodes represented in polar form is preferable to study the geometry of noncircular gears. The orientation of the tangent to the centrode is determined by the angle μ formed by the position vector to the centrode and the tangent (see Section 2.5).

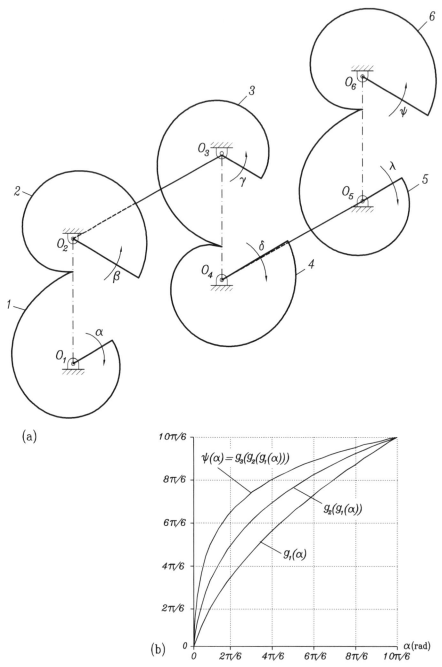

(a)

(b)

Figure 1.2.13. Illustration of a gear drive formed by three pairs of noncircular gears for the generation of given function $\psi_{(\alpha)}$.

CURVATURE OF CENTRODES. Determination of the curvature of centrodes of noncircular gears is necessary for the choice of the generating tool (rack cutter, hob, or shaper) that can be applied for the generation of noncircular gears and for the avoidance of undercutting. Equations for the determination of the curvature of a polar curve are presented in Section 2.8 and the conditions for the observation of convexity of centrodes are formulated.

EVOLUTE AND INVOLUTE OF PROFILES OF TEETH. The evolute for profiles of a tooth is determined as the sequence of centers of curvature determined for the points of the profiles of the tooth. Each profile of the tooth of a noncircular gear has its evolute. Profiles of both sides of a tooth have different evolutes. The evolute for all profiles of the teeth of the same circular gear is the same curve and is called the *base circle*.

The concept of evolute may be described geometrically as follows: (i) determine at each point of the tooth profile the normal to the profile – the sequence of normals will form a family of straight lines; (ii) the envelope to the family of normals is the evolute.

The tooth profile is a curve that may be called *involute*. Tooth profiles of the same noncircular gear are represented by different curves (different involutes).

IDENTITY OF MATING CENTRODES. The generation of a function is performed by the meshing of two conjugated centrodes and, generally, the mating centrodes are not identical. However, in some particular cases the mating centrodes might be identical by observation of certain requirements for the transmission function of mating centrodes (see Section 4.5).

MATRIX APPROACH FOR DERIVATION OF EQUATION OF MESHING. The gear tooth surface Σ_g generated by the tool surface Σ_t is determined as the envelope to the family of tool surfaces in a coordinate system fixed to the gear (Litvin & Fuentes, 2004). The applied procedure requires determination of the equation of meshing of surfaces Σ_g and Σ_t.

The most effective way to derive the equation of meshing is based on the application of the scalar product

$$\mathbf{r}_t(u, \theta) \cdot \mathbf{v}_r^{(t,g)} = f(u, \theta, \phi) = 0 \qquad (1.3.1)$$

where $\mathbf{r}_t(u, \theta)$ is the tool surface represented in two-parameter form (by parameters (u, θ)) and $\mathbf{v}_r^{(t,g)}$ is the velocity in relative motion. Determination of Eq. (1.3.1) has been based on differentiation of the matrices that describe the coordinate transformation (see Section 5.2).

AVOIDANCE OF UNDERCUTTING. The proposed approach is based on representation of profiles of teeth of a noncircular gear by a set of profiles of teeth of circular gears (see Section 4.3.2.2).

2 Centrodes of Noncircular Gears

2.1 Introduction

Noncircular gears are mainly designed for the transformation of rotation between parallel axes with a constant center distance performed with variation of ratio of angular velocities $\omega^{(1)}/\omega^{(2)}$.

Figure 2.1.1 shows schematically the transformation of rotation of centrodes of noncircular gears. Planes Π_1 and Π_2 (Fig. 2.1.1(a)) are perpendicular to axes $(\overline{a-a})$ and $(\overline{b-b})$ (Fig. 2.1.1(b)), respectively, of gear rotation. The rotation of gear 1 is performed counterclockwise as shown by arc a and rotation of gear 2 is performed clockwise as shown by arc b (Fig. 2.1.1(a)). Vector $\mathbf{v}^{(i)}(i = 1, 2)$ of linear velocity of rotation of a point of gear belongs to plane Π_i. Vectors $\boldsymbol{\omega}^{(1)}$ and $\boldsymbol{\omega}^{(2)}$ of angular velocities of gear rotation are directed along the axes of rotation (Fig. 2.1.1(b)).

The centrodes are planar curves that roll over each other in the process of transformation of rotation. Figure 2.1.2 shows an example of centrodes of elliptical noncircular gears. The centrodes contact each other in the process of transformation of rotation at a point I that moves along $O_1 - O_2$ (Fig. 2.1.1(a)). Vectors $\mathbf{v}^{(1)}$ and $\mathbf{v}^{(2)}$ represent the linear velocities of rotation about O_1 and O_2, respectively, and are determined as

$$\mathbf{v}^{(1)} = \boldsymbol{\omega}^{(1)} \times \overline{O_1 I}, \qquad \mathbf{v}^{(2)} = \boldsymbol{\omega}^{(2)} \times \overline{O_2 I} \qquad (2.1.1)$$

The condition of rolling of centrodes is observed by the vector equation

$$\mathbf{v}^{(1)} = \mathbf{v}^{(2)} \qquad (2.1.2)$$

Equation (2.1.2) shows that the relative velocity at the point of tangency of centrodes is equal to zero. Thus, the condition of rolling of centrodes is observed by the vector equation

$$\mathbf{v}^{(12)} = \mathbf{v}^{(1)} - \mathbf{v}^{(2)} = \mathbf{0} \qquad (2.1.3)$$

Similarly, we have

$$\mathbf{v}^{(21)} = \mathbf{v}^{(2)} - \mathbf{v}^{(1)} = \mathbf{0} \qquad (2.1.4)$$

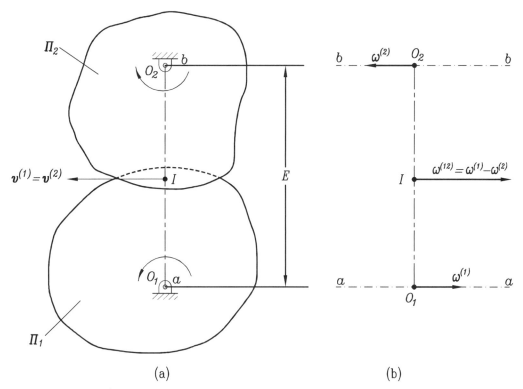

Figure 2.1.1. Illustration of transformation of rotation: (a) O_1 and O_2 are the centers of rotation; (b) vectors $\boldsymbol{\omega}^{(1)}$ and $\boldsymbol{\omega}^{(2)}$ of angular velocities are directed along the respective axes of rotation.

Figure 2.1.2. Illustration of mating centrodes that roll over each other and contact each other at point I, which is moving along center distance $O_1 - O_2$.

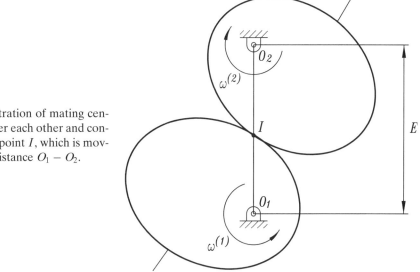

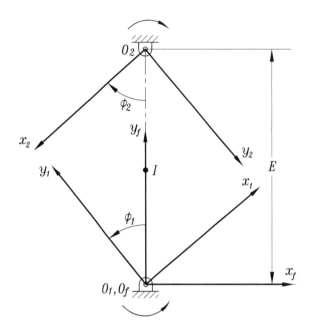

Figure 2.2.1. Illustration of fixed coordinate system $S_f(x_f, y_f)$ and movable coordinate systems $S_1(x_1, y_1)$, $S_2(x_2, y_2)$.

Point I of the tangency of centrodes is the instantaneous center of rotation in relative motion. Considering that one of the centrodes is held at rest, for instance centrode 2, then we may consider that centrode 1 is rotated about centrode 2 at point I (Fig. 2.1.1(b)) with angular velocity $|\boldsymbol{\omega}^{(12)}| = |\boldsymbol{\omega}^{(1)}| + |\boldsymbol{\omega}^{(2)}|$.

2.2 Centrode as the Trajectory of the Instantaneous Center of Rotation

The determination of centrodes is the first step in the design of noncircular gears, and for this purpose we apply: (i) two movable coordinate systems $S_1(x_1, y_1)$ and $S_2(x_2, y_2)$ rigidly connected with gears 1 and 2, and (ii) a fixed coordinate system $S_f(x_f, y_f)$ rigidly connected to the frame of the gear drive (see Figure 2.2.1).

During the transformation of rotation of noncircular gears 1 and 2, the instantaneous center of rotation I moves along center distance E, and the trajectory traced out by the instantaneous center of rotation in S_f is represented as the straight line $\mathbf{r}_f(u)$ where $u = |\overline{O_1 I}|$.

Simultaneously, the instantaneous center of rotation moves in coordinate systems $S_1(x_1, y_1)$ and $S_2(x_2, y_2)$ along centrodes σ_1 and σ_2 of gears 1 and 2.

We may determine the centrodes as trajectories traced out by point I in coordinate systems S_1 and S_2, respectively, and represented as

$$\mathbf{r}_1(u, \phi_1) = \mathbf{M}_{1f}(\phi_1)\mathbf{r}_f(u) \tag{2.2.1}$$

$$\mathbf{r}_2(u, \phi_2) = \mathbf{M}_{2f}(\phi_2)\mathbf{r}_f(u) \tag{2.2.2}$$

Matrices $\mathbf{M}_{if}(i = 1, 2)$ represent the coordinate transformation from S_f to S_i ($i = 1, 2$).

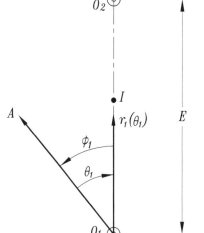

Figure 2.3.1. Illustration of polar axis $\overline{O_1 A}$, center distance E, rotation angle ϕ_1, and parameter θ_1.

2.3 Concept of Polar Curve

The centrodes are determined as polar curves. Figure 2.3.1 shows (i) the polar axis $\overline{O_1 A}$ of a driving centrode that passes through its center of rotation O_1; (ii) $E = |\overline{O_1 O_2}|$ is the shortest center distance between centrodes 1 and 2, where O_2 is the center of rotation of centrode 2; (iii) $r_1(\theta_1)$ is the position vector of centrode 1, where θ_1 is the angle formed by polar axis $\overline{O_1 A}$ and the center distance $|\overline{O_1 O_2}|$; parameter θ_1 is measured clockwise, and function $r_1(\theta_1)$ determines centrode 1 as a polar curve; (iv) ϕ_1 is the angle of rotation of centrode 1 measured counterclockwise ($\phi_1 = \theta_1$).

2.4 Derivation of Centrodes

We consider the following cases of the derivation of centrodes (i) wherein the derivative function $m_{12}(\phi_1) = d\phi_1/d\phi_2$ and the center distance of the gear drive with noncircular gears are given, and (ii) wherein function $y(x)$, $x_1 \leq x \leq x_2$, which must be generated by the mechanism of noncircular gears, is given.

CASE 1: DERIVATION OF CENTRODES CONSIDERING AS GIVEN FUNCTION $m_{12}(\phi_1)$

Step 1. Taking into account that

$$m_{12}(\phi_1) = \frac{d\phi_1}{d\phi_2} = \frac{r_2}{r_1} = \frac{E - r_1}{r_1} \qquad (2.4.1)$$

we obtain the following expression for centrode 1 (centrode σ_1):

$$r_1(\phi_1) = \frac{E}{1 + m_{12}(\phi_1)} \qquad (2.4.2)$$

Taking into account that $\phi_1 = \theta_1$, we obtain the polar equation of σ_1

$$r_1(\theta_1) = \frac{E}{1 + m_{12}(\theta_1)} \tag{2.4.3}$$

Step 2. Centrode 2 (centrode σ_2) is determined by the equations

$$r_2(\phi_1) = E \frac{m_{12}(\phi_1)}{1 + m_{12}(\phi_1)} \tag{2.4.4}$$

$$\phi_2(\phi_1) = \int_0^{\phi_1} \frac{d\phi_1}{m_{12}(\phi_1)} \tag{2.4.5}$$

The derivative function $m_{12}(\phi_1)$ must be a smooth, differentiable, and positive function. The requirement of positive values of $m_{12}(\phi_1)$ in the area $\phi_1^{(1)} \leq \phi \leq \phi_1^{(2)}$ must be observed because the rotation of the gears may be performed in one direction only as a continued one (see Section 10.3 on generation of non-monotonous functions). Variation of the magnitude of $m_{12}(\phi_1)$ may be accompanied with an unfavorable magnitude of the pressure angle, which may require the design of a mechanism with more than one pair of mating gears.

CASE 2: DERIVATION OF CENTRODES CONSIDERING AS GIVEN FUNCTION $y(x)$, $x_1 \leq x \leq x_2$

Step 1. The angles of rotation of the driving and driven gears are proportional to x and $f(x)$, respectively. We then obtain the following equations for the scalar coefficients k_1 and k_2:

$$\phi_1 = k_1(x - x_1) \tag{2.4.6}$$

$$\phi_2 = k_2 \big[f(x) - f(x_1) \big] \tag{2.4.7}$$

Step 2. The gear ratio is determined as

$$m_{12}(x) = \frac{d\phi_1}{d\phi_2} = \frac{k_1}{k_2 f'(x)} \tag{2.4.8}$$

The following requirements must be observed:

 (i) The derivative function $m_{12}(x)$, $x_1 \leq x \leq x_2$, must be positive.
(ii) The pressure angle (see Section 4.3.2.2) should not exceed the limiting values.

Such requirements may be fulfilled by:

(a) Application of multithread noncircular gears wherein $\phi_{1max} \geq 2\pi$ (see Fig. 1.2.4).
(b) Use of a gear mechanism with more than one pair of gears (see Section 10.5).

2.5 Tangent to Polar Curve

Figure 2.5.1 shows polar axis $\overline{OA}$ and two points M_0 and M_2 of the polar curve infinitesimally close to each other. Vector **t** is the tangent to the polar curve at point M_0.

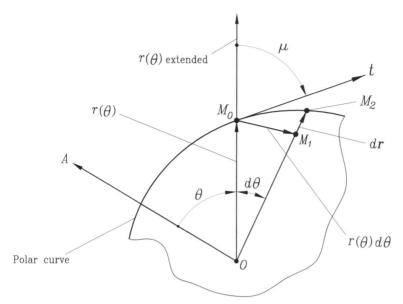

Figure 2.5.1. For derivation of angle μ formed by extended position vector $r(\theta)$ and tangent **t** to polar curve.

Taking into account that $|\widehat{M_0 M_1}| = r(\theta)d\theta$ and segment $|\overline{M_1 M_2}| = |d\mathbf{r}|$, we find that

$$\tan \mu = \frac{r}{dr/d\theta} \tag{2.5.1}$$

Angle μ (Fig. 2.5.1) is formed by extended position vector $\overline{OM_0}$ and tangent **t**. Equation (2.5.1) is the basic equation for the determination of tangent **t** to polar curve. Tangent **t** to a polar curve is directed in accordance to measurements of polar angle θ (Fig. 2.5.1). Figure 2.5.2(a) shows two mating centrodes σ_1 and σ_2 that contact each other at the instantaneous center of rotation I. The common tangent to the centrodes is **t**. Parameters ϕ_i and θ_i $(i = 1, 2)$ represent the angles of rotation of centrodes and scalar parameters.

Our goal is to prove that angles μ_1 and μ_2 (Figs. 2.5.2(b) and 2.5.2(c)) are related at point I of tangency of centrodes σ_1 and σ_2 as

$$\mu_1 + \mu_2 = \pi \tag{2.5.2}$$

(i) Centrode σ_1 is represented by Eq. (2.4.3). Following Eq. (2.5.1), we obtain

$$\tan \mu_1 = \frac{r_1(\theta_1)}{dr_1/d\theta_1} = -\frac{1 + m_{12}(\theta_1)}{dm_{12}(\theta_1)/d\theta_1} \tag{2.5.3}$$

(ii) Centrode σ_2 is represented by Eqs. (2.4.4) and (2.4.5). Taking into account that $\phi_1 = \theta_1$, we obtain $\tan \mu_2 = -\tan \mu_1$, and Eq. (2.5.2) is observed.

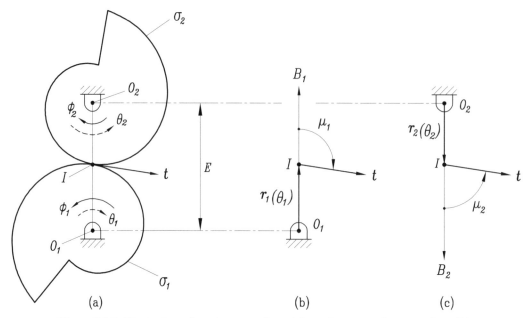

Figure 2.5.2. Illustration of angles μ_1 and μ_2: (a) mating centrodes σ_1 and σ_2; (b) measurement of angle μ_1; (c) measurement of angle μ_2; $O_iB_i(i=1,2)$ shows the extended position vector $r_i(\theta_i)$.

2.6 Conditions for Design of Centrodes as Closed Form Curves

Mechanisms applied in automatic machines are designed for continuous motion, and the centrodes must be designed as closed curves. The conditions of obtaining centrodes as closed curves are formulated below (i) for a conventional mechanism designed for transformation of motions, and (ii) for a mechanism designed for generation of a given function. Conditions of obtaining closed curves are formulated for the driving and driven gear separately.

DESIGN OF CENTRODE 1. We recall that centrode $r_1(\theta_1)$ $(\theta_1 = \phi_1)$ is represented by Eq. (2.4.2). This equation shows that centrode 1 may be obtained as a closed form curve if $m_{12}(\phi_1)$ is a periodic function with period $T = T_1/n_1$, where T_1 is the period of revolution of gear 1 and n_1 is an integer number.

DESIGN OF CENTRODE 2. The centrode is determined by Eqs. (2.4.4) and (2.4.5). Equation (2.4.4) yields that centrode 2, $r_2(\phi_1)$, is a closed form curve if $m_{12}(\phi_1)$ is a periodic function with the period $T = T_1/n_1$. This means that the interval of variation of ϕ_1 during one revolution is a multiple of $2\pi/n_1$.

In addition, the following requirements must be observed:

(i) The ratio between the revolutions n_1 and n_2 of centrodes 1 and 2, respectively, must be

$$\frac{T_1}{n_1} = \frac{T_2}{n_2} \tag{2.6.1}$$

where n_1 and n_2 are integer numbers.

(ii) The gear center distance E must be determined as a function of n_1 and n_2 as follows:

(a) We represent $r_2(\phi_1)$ by the equation

$$r_2(\phi_1) = E - r_1(\phi_1) \tag{2.6.2}$$

(b) The angle ϕ_2 of rotation of gear 2 is determined as

$$\phi_2(\phi_1) = \int_0^{\phi_1} \frac{d\phi_1}{m_{12}(\phi_1)} = \int_0^{\phi_1} \frac{r_1(\phi_1)}{r_2(\phi_1)} d\phi_1 = \int_0^{\phi_1} \frac{r_1(\phi_1)}{E - r_1(\phi_1)} d\phi_1 \tag{2.6.3}$$

Observation of Eqs. (2.6.1) and (2.6.3) yields the following equation for the determination of center distance E:

$$\phi_2 = \frac{2\pi}{n_2} = \int_0^{\frac{2\pi}{n_1}} \frac{r_1(\phi_1)}{E - r_1(\phi_1)} d\phi_1 \tag{2.6.4}$$

The design of gears with closed form centrodes requires observation of Eqs. (2.6.1) and (2.6.4) and is discussed in Chapter 4 for conventional and modified elliptical gears, in Chapter 6 for eccentric involute gear drives, and in Chapter 7 for internal gear drives.

Problem 2.6.1 and its solution are provided for the comprehension of approaches applied for the design of noncircular gear drives.

PROBLEM 2.6.1. TOWARD DESIGN OF A GEAR DRIVE WITH ELLIPTICAL PINION. The design of centrodes of gear drives with conventional and modified elliptical centrodes may be performed analytically by application of the following integrals represented by Dwight (Dwight, 1961):

$$\int_0^\pi \frac{dx}{1 \pm a \cos x} = \frac{\pi}{\sqrt{1 - a^2}} \tag{2.6.5}$$

$$\int \frac{dx}{a + b \cos x} = \frac{2}{\sqrt{a^2 - b^2}} \arctan \frac{(a - b) \tan \frac{x}{2}}{\sqrt{a^2 - b^2}} \qquad [a^2 > b^2] \tag{2.6.6}$$

Step 1. Centrode σ_1.

Centrode σ_1 is a conventional ellipse, and the center of rotation is the lower focus of the ellipse. The polar equation of centrode σ_1 is then (see Section 4.2)

$$r_1(\phi_1) = \frac{p}{1 - e \cos \phi_1} = \frac{a(1 - e^2)}{1 - e \cos \phi_1}, \qquad 0 \le \phi_1 \le 2\pi n \tag{2.6.7}$$

Here, a is the larger axis of the ellipse, $e = c/a$ is the eccentricity of the ellipse with $2c$ being the distance between the focuses of the ellipse, n is the number of revolutions that centrode σ_1 performs for one revolution of conjugated centrode σ_2, and ϕ_1 is the angle of rotation of σ_1 (Fig. 2.6.1). The centrode σ_1 is already a closed form curve.

Figure 2.6.1 shows that center O_1 of rotation of centrode 1 is located in the lower focus of the ellipse; O is the center of the ellipse; $\overline{O_1 A_1}$ is the polar axis for

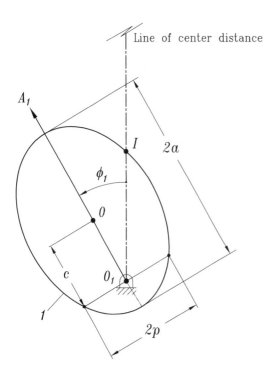

Line of center distance

A_1

I $2a$

ϕ_1

0

c 0_1

1

$2p$

Figure 2.6.1. Illustration of geometric parameters applied in Eq. (2.6.7).

representation of the ellipse as a polar curve; and $|O_1 I| = r_1(\phi_1)$ is the position vector of the polar curve.

Determination of centrode σ_2 as a closed form curve is discussed in the following.

Step 2. Derivative function $m_{21}(\phi_1)$.

The derivation of $m_{21}(\phi_1)$ is based on application of Eqs. (2.6.7) and (2.4.1), which yield

$$m_{21}(\phi_1) = \frac{1}{m_{12}(\phi_1)} = \frac{p}{E - p - Ee \cos \phi_1}$$

$$= \left(\frac{p}{E - p}\right) \frac{1}{1 - q \cos \phi_1}, \quad q = \frac{Ee}{E - p} \qquad (2.6.8)$$

Step 3. Derivation of center distance E.

The derivation of E is based on the Eqs. (2.6.1) and (2.6.3), which yield

$$\phi_2(\phi_1)_{\phi_1=2\pi} = 2\pi = n \int_0^{2\pi} \frac{p}{E - p - Ee \cos \phi_1} d\phi_1$$

$$= \left(\frac{pn}{E - p}\right) \int_0^{2\pi} \frac{d\phi_1}{1 - q \cos \phi_1} = \frac{2pn}{E - p} \int_0^{\pi} \frac{d\phi_1}{1 - q \cos \phi_1} \qquad (2.6.9)$$

where n is the number of revolutions of pinion centrode σ_1 performed for one revolution of gear centrode σ_2. By applying Eq. (2.6.5), Eq. (2.6.9) yields the following quadratic equation:

$$E^2(1 - e^2) - 2Ep + p^2(1 - n^2) = 0 \qquad (2.6.10)$$

The solution of Eq. (2.6.10) for E is

$$E = a \left[1 + \sqrt{1 + (1 - e^2)(n^2 - 1)} \right] \qquad (2.6.11)$$

The importance of Eq. (2.6.11) is that it works for conventional elliptic gear drives, modified elliptical gear drives, oval gears, and for gears with lobes (see Chapter 4).

Step 4. Centrode σ_2 of driven gear.

Centrode σ_2 is represented by the following equations:

$$r_2(\phi_1) = E - r_1(\phi_1) = E - \frac{a(1 - e^2)}{1 - e \cos \phi_1} \qquad (2.6.12)$$

$$\phi_2(\phi_1) = \int_0^{\phi_1} m_{21}(\phi_1) d\phi_1 = \left(\frac{p}{E - p} \right) \int_0^{\phi_1} \frac{d\phi_1}{1 - q \cos \phi_1} \qquad (2.6.13)$$

The transmission function (2.6.9) may be determined analytically using Eq. (2.6.6). We then obtain the following expression for $\phi_2(\phi_1)$:

$$\tan \frac{\phi_2}{2} = \left(\frac{1 + e}{1 - e} \right) \tan \frac{\phi_1}{2} \qquad (2.6.14)$$

2.7 Observation of Closed Centrodes for Function Generation

Basic ideas of the application of noncircular gears for generation of the function of $y(x)$, $x_1 \leq x \leq x_2$, are formulated by Eqs. (2.4.6) to (2.4.8). The following conditions must be observed to obtain closed form centrodes for generation of the function in addition to the previously mentioned equations:

(a) The derivative function $y'(x)$ must be periodic and of period $\dfrac{x_2 - x_1}{n_1}$, where n_1 is an integer number.

(b) The scalar coefficients k_1 and k_2 of Eqs. (2.4.6) and (2.4.7) must satisfy the following relations:

$$k_1 = \frac{2\pi}{x_2 - x_1} \qquad (2.7.1)$$

$$k_2 = \frac{2\pi}{f(x_2) - f(x_1)} \qquad (2.7.2)$$

2.8 Basic and Alternative Equations of Curvature of Polar Curve

Determination of the curvature of centrodes of noncircular gears is required for the solution of the following problems:

(i) Determination of the type of generating tool (rack cutter, hob, or shaper) that can be applied for generation of noncircular gears. A rack cutter or a hob may be applied only for convex centrodes.

(ii) Avoidance of undercutting (see Section 4.3.2.2).

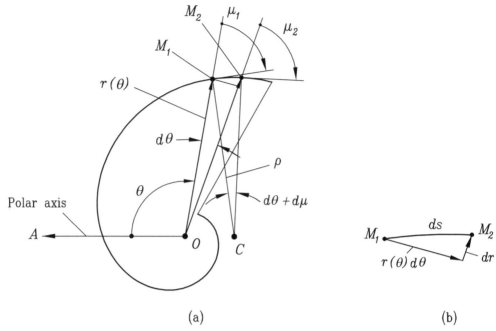

(a)   (b)

Figure 2.8.1. For derivation of the curvature of polar curve: (a) $\overline{OA}$ is the polar axis, $r(\theta)$ is the position vector, ρ is the curvature radius, and C is the curvature center; (b) $ds = \widehat{M_1 M_2}$, and $r(\theta)d\theta$ and dr are the arc components of $\widehat{M_1 M_2}$,

Figure 2.8.1(a) shows a polar curve wherein $\overline{CM_1}$ and $\overline{CM_2}$ are the normals to the curve that form angle $(d\theta + d\mu)$ and C is the curvature center. The length ds of the small arc $\widehat{M_1 M_2}$ is determined as (Fig. 2.8.1(b))

$$ds = \sqrt{[r(\theta)d\theta]^2 + (dr)^2} = r(\theta)\sqrt{1 + \left(\frac{dr/d\theta}{r(\theta)}\right)^2}\,d\theta$$

$$= r(\theta)\sqrt{1 + \frac{1}{\tan^2 \mu}}\,d\theta = \frac{r(\theta)}{\sin \mu}\,d\theta \qquad (2.8.1)$$

Considering $\widehat{M_1 M_2}$ as the small arc of radius ρ, we represent ds as

$$ds = \rho(d\theta + d\mu) \qquad (2.8.2)$$

Equations (2.8.1) and (2.8.2) yield

$$\rho(\theta) = \frac{r(\theta)}{\sin \mu}\left(\frac{d\theta}{d\theta + d\mu}\right) \qquad (2.8.3)$$

Determination of function $d\mu/d\theta = f(\theta)$ is based on the following considerations:

(i) We recall that $\tan \mu = \dfrac{r}{dr/d\theta}$ (see Eq. (2.5.1)). After differentiation, we obtain

$$d\mu = \frac{\cos^2 \mu \left[\left(\dfrac{dr}{d\theta} \right)^2 - r \dfrac{d^2 r}{d\theta^2} \right]}{\left(\dfrac{dr}{d\theta} \right)^2} d\theta \qquad (2.8.4)$$

(ii) Continuing transformations, we obtain

$$d\mu + d\theta = \frac{\cos^2 \mu \left[\left(\dfrac{dr}{d\theta} \right)^2 - r \dfrac{d^2 r}{d\theta^2} \right] + \left(\dfrac{dr}{d\theta} \right)^2}{\left(\dfrac{dr}{d\theta} \right)^2} d\theta \qquad (2.8.5)$$

(iii) Equations (2.8.5) and (2.5.1) yield

$$d\mu + d\theta = \frac{r(\theta)(2 - \sin^2 \mu) - \sin^2 \mu \dfrac{d^2 r}{d\theta^2}}{\left(\dfrac{dr}{d\theta} \right)^2} d\theta \qquad (2.8.6)$$

Equations (2.8.6) and (2.8.3) yield the following expression for the curvature radius:

$$\rho(\theta) = \frac{[r(\theta)]^2}{\sin \mu \left[r(\theta)(2 - \sin^2 \mu) - \sin^2 \mu \dfrac{d^2 r}{d\theta^2} \right]} \qquad (2.8.7)$$

The curvature radius $\rho(\theta)$ is represented as well in differential geometry (Korn & Korn, 1968) as

$$\rho(\theta) = \frac{\left[r(\theta)^2 + \left(\dfrac{dr}{d\theta} \right)^2 \right]^{3/2}}{r(\theta)^2 + 2 \left(\dfrac{dr}{d\theta} \right)^2 - r(\theta) \dfrac{d^2 r}{d\theta^2}} \qquad (2.8.8)$$

We may obtain (2.8.8) from (2.8.7) taking into account that

$$r(\theta)^2 + \left(\frac{dr}{d\theta} \right)^2 = r(\theta)^2 + \frac{r(\theta)^2}{\tan^2 \mu} = \frac{r(\theta)^2}{\sin^2 \mu} \qquad (2.8.9)$$

$$2 \left(\frac{dr}{d\theta} \right)^2 = 2 \frac{r(\theta)^2}{\tan^2 \mu} \qquad (2.8.10)$$

2.9 Conditions of Centrode Convexity

Equation (2.8.8) of curvature radius ρ yields the following condition of centrode convexity ($\rho > 0$):

$$r(\phi)^2 + 2\left(\frac{dr}{d\phi}\right)^2 - r(\theta)\frac{d^2r}{d\phi^2} > 0 \qquad (2.9.1)$$

We recall that $\phi = \theta$ (see Fig. 2.3.1). Using equations of centrodes and inequality (2.9.1), we obtain the conditions for centrodes convexity.

Equation (2.4.2) and inequality (2.9.1) show that centrode 1 is convex by observation of the condition

$$1 + m_{12}(\phi_1) + m_{12}''(\phi_1) \geq 0 \qquad (2.9.2)$$

Similarly, using Eqs. (2.4.4) and (2.4.5) for centrode 2 and inequality (2.9.1), we obtain conditions of centrode 2 convexity as

$$1 + m_{12}(\phi_1) + [m_{12}'(\phi_1)]^2 - m_{12}(\phi_1)m_{12}''(\phi_1) \geq 0 \qquad (2.9.3)$$

Here,

$$m_{12}'(\phi_1) = \frac{d}{d\phi_1}\left(m_{12}(\phi_1)\right); \qquad m_{12}''(\phi_1) = \frac{d^2}{d^2\phi_1}\left(m_{12}(\phi_1)\right) \qquad (2.9.4)$$

When the inequalities (2.9.2) and (2.9.3) turn into equalities, there is a centrode point where $\rho = \infty$.

Using the approach discussed here for the generation of function by noncircular gears, we obtain the following conditions of convexity of the centrodes of the driving and driven gears, respectively.

Let us consider now the case that the noncircular gears are designed for generation of the function $z = f(u)$. Then we obtain

$$m_{12}(u) = \frac{k_1}{k_2\, f'(u)} \qquad (2.9.5)$$

where the relation between ϕ_1 and u is

$$\phi_1 = k_1(u - u_1) \qquad (2.9.6)$$

Respectively, we obtain

$$m_{12}'(\phi_1) = \frac{d}{d\phi_1}\left(m_{12}(\phi_1)\right) = \frac{d}{du}\left(m_{12}(\phi_1)\right)\frac{du}{d\phi_1} = -\left(\frac{1}{k_2}\right)\frac{f''(u)}{[f'(u)]^2} \qquad (2.9.7)$$

$$m_{12}''(\phi_1) = \left(\frac{1}{k_1 k_2}\right)\frac{2[f''(u)]^2 - f'(u)f'''(u)}{[f'(u)]^3} \qquad (2.9.8)$$

Finally, the conditions of convexity of centrodes 1 and 2 are represented respectively as

$$k_1 k_2[f'(u)]^3 + k_1^2[f'(u)]^2 + 2[f''(u)]^2 - f'''(u)f'(u) \geq 0 \qquad (2.9.9)$$

$$k_2[f'(u)]^3[k_1 + k_2 f'(u)] + f'(u)f'''(u) - [f''(u)]^2 \geq 0 \qquad (2.9.10)$$

3 Evolutes and Involutes

3.1 Introduction and Terminology

CENTRODES AND PITCH CURVES. *Centrodes* (see Chapter 2) may be considered as frictional disks (circular or noncircular) that roll over each other in the process of meshing. The centrodes are provided with teeth, and such a mechanism (gear mechanism) transforms rotation by the meshing of teeth.

Figure 3.1.1 shows the centrodes (as mating circles) of a gear mechanism with circular gears. For the purpose of simplification of drawings, only two profiles of teeth are shown in Fig. 3.1.1. The instantaneous point of contact of the tooth profiles is shown in Fig. 3.1.1 as I.

Points $M_i^{(1)}$ and $M_i^{(2)}$ in Fig. 3.1.1 represent points of intersection of neighboring profiles with the centrodes of radii $r_p^{(1)}$ and $r_p^{(2)}$. The distance p between two neighboring points $M_i^{(j)}$ ($j = 1, 2$) is measured along the arc of the respective centrode and is determined as $p = m\pi$, where m is the module of the gears.

It will be shown here that even neighboring profiles may have different evolutes (for instance, see Fig. 3.1.3). There is an exemption with involute spur gears, wherein the evolutes of neighboring profiles are arcs of the same circles called *base circles* of radii $r_b^{(i)}$ ($i = 1, 2$). Centrodes of radii $r_p^{(i)}$ ($i = 1, 2$) are usually called *pitch circles*, but the preferable notation is centrodes.

CONCEPT OF EVOLUTE AND INVOLUTE OF PLANAR CURVES. The concept of evolute and involute of a planar curve has been developed in differential geometry.

We call *involute* the tooth profile obtained in the transverse section of the gear tooth. The profile might be represented by a curve or, in some cases, by a straight line. Figure 3.1.2 shows the tooth profile as a planar curve σ_1 formed by points $M_1, M_2, \cdots, M_n$. We may determine for each point M_i of σ_1 the curvature center N_i of involute σ_1 at point M_i of the involute. Straight lines $\overline{M_i N_i}$ ($i = 1, 2, \cdots, n$) are the normals to the involute. Following the definition in differential geometry, we determine the evolute σ_2 of curve σ_1 as the locus of curvature centers of σ_1 (Fig. 3.1.2). Thus, points N_i ($i = 1, 2, \cdots, n$) of normals $M_i - N_i$ to σ_1 form curve

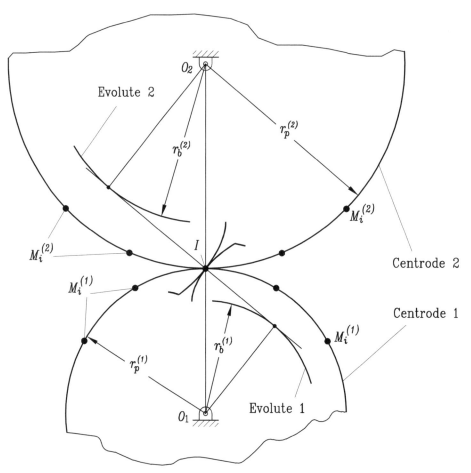

Figure 3.1.1. Illustration of centrodes of radii r_{p_1} and r_{p_2}, evolutes of radii r_{b_1} and r_{b_2}, and points $M_i^{(1)}$ and $M_i^{(2)}$ of intersection with centrodes of neighboring tooth profiles.

$N_1 - N_2 - \cdots - N_n$ (Fig. 3.1.2), which is called the *evolute*. Straight lines $\overline{M_i N_i}$, the normals to involute σ_1, are the tangents to the evolute (curve σ_2, Fig. 3.1.2).

We may also interpret evolute σ_2 as the envelope to the family of normals $M_i N_i$ to curve σ_1. Considering as given evolute σ_2, we may trace out curve σ_1 by a point of a string as it unwinds from σ_2.

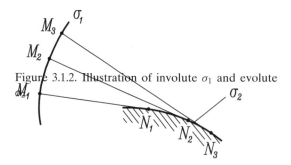

Figure 3.1.2. Illustration of involute σ_1 and evolute σ_2

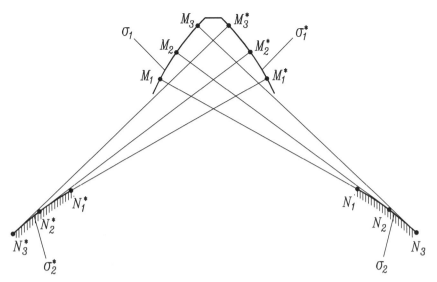

Figure 3.1.3. Illustration of involutes (σ_1, σ_1^*) and evolutes (σ_2, σ_2^*) for both profiles of gear tooth.

APPLICATION OF CONCEPT OF EVOLUTE AND INVOLUTE IN THE THEORY OF GEARING.
A transverse section of a single gear tooth provides two involute profiles and two evolutes (Fig. 3.1.3). The exemption of this rule is the so-called involute gearing, wherein the envelope to normals $M_i - N_i$ and $M_i^* - N_i^*$ to the tooth profiles σ_1 and σ_1^* is the same circle, called the *base circle* (see Fig. 3.1.1).

Figure 3.1.3 shows in exaggerated style the tooth of an involute spur gear. Points $M_1 - M_2 - \cdots - M_n$ form the involute profile of the left side of the tooth. Similarly, points $M_1^* - M_2^* - \cdots - M_n^*$ form the right side of the tooth. Normals $M_i - N_i$ $(i = 1, \cdots, n)$ to the involute profile σ_1 form a family of straight lines that are tangents to evolute σ_2 (Fig. 3.1.3). Points N_i form evolute σ_2 as $N_1 - N_2 - \cdots - N_n$. Respectively, normals $M_i^* - N_i^*$ $(i = 1, \cdots, n)$ to profile σ_1^* are tangents to evolute σ_2^*. Thus, points $N_1^* - N_2^* - \cdots - N_n^*$ form envelope σ_2^*.

As mentioned previously, in the case of an involute spur gear the evolutes σ_2 and σ_2^* for both involute profiles σ_1 and σ_1^* are arcs of the same circle, the base circle. For the case of noncircular gears, profiles σ_1 and σ_1^* of the tooth have different evolutes.

3.2 Determination of Evolutes

In the process of generation of noncircular gears, centrode 1 of the tool (rack cutter, shaper) is rolling over centrode 2 of the noncircular gear being generated (Fig. 3.2.1). Point P is the common point of tangency of centrodes 1 and 2. We emphasize that the normal to the tooth profile and the common tangent $\mathbf{t}$ to the centrode (1 or 2) forms a constant angle α_c that is the profile angle of the tool.

The position vector $r(\theta)$ of centrode 2 (of the noncircular gear) and the normal to the tooth profile form angle λ_i $(i = l, r)$, wherein subindex i of λ_i indicates the normals to the left $(i = l)$ and right $(i = r)$ sides of profiles of the noncircular gear.

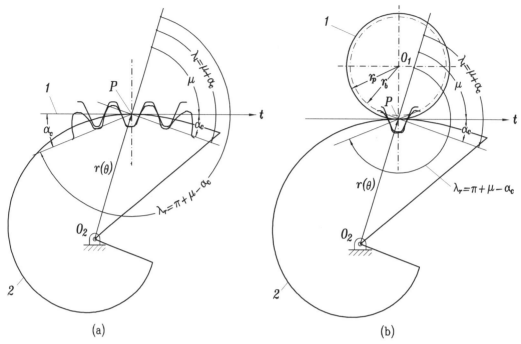

Figure 3.2.1. Illustration of orientation of normals to the tooth profile by angle λ (a) as generated by a rack cutter; (b) as generated by a shaper.

It follows from the drawings that

$$\lambda_l = \mu + \alpha_c \qquad\qquad (3.2.1)$$

$$\lambda_r = \pi + \mu - \alpha_c \qquad\qquad (3.2.2)$$

where α_c is the profile angle of the tool.

Figure 3.2.2 shows polar curve σ_1, the centrode of a noncircular gear. For the purpose of simplification of notation, we are not using the subindexes of notation of λ, μ, or $r(\theta)$, but the obtained equations for the evolutes may be applied for evolutes of left and right profiles.

The analytical determination of evolute σ_2 requires determination of $l = \overline{MC}$, where C is the point of tangency of normal $\mathbf{N}$ with the evolute. Knowing parameter λ – see Eqs. (3.2.1) and (3.2.2) – and parameter $l = |\overline{MC}|$, it becomes possible to determine the evolute of profiles.

Determination of parameter $l = |\overline{MC}|$ (Fig. 3.2.2) requires application of the theory of envelope to the family of normals $\mathbf{N}$. The illustration of derivations is performed by application of a linkage shown in Fig. 3.2.2 with three movable links. Link 1 is provided with slider 2 and performs rotation about point O, simulating the centrode of the gear. Link 3 may perform rotation about point M, being the tangent to the evolute.

The velocity $\mathbf{v}$ of point C of the normal is considered as the sum of the following components: (i) $\mathbf{v}_e$ in rotation together with link 1 about O; (ii) velocity $\mathbf{v}_r$ in

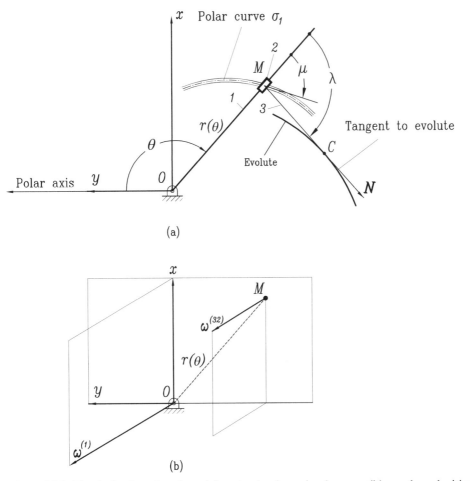

Figure 3.2.2. For derivation of evolute: (a) centrode of noncircular gear; (b) angular velocities of rotation of links 1 and 3 about axis M.

translation with slider 2 along the centrode; (iii) velocity $\mathbf{v}_{32}$ in rotation about point M of slider 2. Velocity $\mathbf{v}_c$ as the sum of the three components is represented by

$$\mathbf{v}_c = \mathbf{v}_e + \mathbf{v}_r + \mathbf{v}_{32} = \boldsymbol{\omega}^{(1)} \times (\mathbf{r} + \mathbf{l}) + \frac{d\mathbf{r}}{dt} + \boldsymbol{\omega}^{(32)} \times \mathbf{l}$$

$$= \boldsymbol{\omega}^{(1)} \times \mathbf{r} + (\boldsymbol{\omega}^{(1)} + \boldsymbol{\omega}^{(32)}) \times \mathbf{l} + \frac{d\mathbf{r}}{dt} \qquad (3.2.3)$$

Here: $\boldsymbol{\omega}^{(1)} = -\dfrac{d\theta}{dt}\mathbf{k}$; $\boldsymbol{\omega}^{(32)} = \dfrac{d\lambda}{dt}\mathbf{k}$; $\mathbf{r} = r\cos\theta\mathbf{i} + r\sin\theta\mathbf{j}$; $\mathbf{l} = l\cos(\theta + \lambda)\mathbf{i} + l\sin(\theta + \lambda)\mathbf{j}$

Taking into account that $\dfrac{r(\theta)}{dr/d\theta} = \tan\mu$ [see Eq. (2.5.1)], we obtain $\dfrac{dr}{dt} = r\cot\mu\dfrac{d\theta}{dt}$ and $\dfrac{d\mathbf{r}}{dt} = -r\cot\mu\dfrac{d\theta}{dt}(\cos\theta\mathbf{i} + \sin\theta\mathbf{j})$, wherein $(\mathbf{i}, \mathbf{j}, \mathbf{k})$ are the unit vectors of coordinate axes (x, y, z) (Fig. 3.2.2).

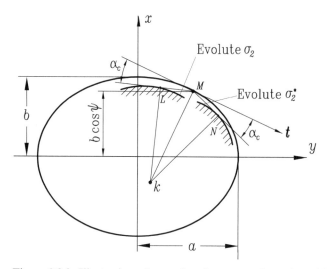

Figure 3.2.3. Illustration of arcs of evolutes σ_2 and σ_2^* of an elliptical centrode.

After transformation, projections of vector velocity $\mathbf{v}_c$ are represented as

$$v_{yc} = r \sin\theta \frac{d\theta}{dt} + l \sin(\theta + \lambda)\left(\frac{d\theta}{dt} + \frac{d\lambda}{d\theta}\right) - r \cot\mu \cos\theta \frac{d\theta}{dt} \qquad (3.2.4)$$

$$v_{xc} = -r \cos\theta \frac{d\theta}{dt} - l \sin(\theta + \lambda)\left(\frac{d\theta}{dt} + \frac{d\lambda}{d\theta}\right) - r \cot\mu \sin\theta \frac{d\theta}{dt} \qquad (3.2.5)$$

Vectors $\mathbf{v}_c$ and $\mathbf{l}$ are collinear, and therefore

$$\frac{v_{yc}}{v_{xc}} = \frac{l_y}{l_x} = \frac{\cos(\theta + \lambda)}{\sin(\theta + \lambda)} \qquad (3.2.6)$$

Equations (3.2.4), (3.2.5), and (3.2.6) yield

$$l(\theta) = r(\theta)\frac{\sin(\lambda - \mu)}{\left(1 + \frac{d\lambda}{d\theta}\right)\sin\mu} \qquad (3.2.7)$$

Equation (3.2.7) may be applied for the determination of evolute of profiles for planar gears with various geometries.

Taking into account Eq. (2.8.7) of curvature radius ρ for planar gears, we obtain after transformations the following expression for the determination of evolute:

$$l(\theta) = \rho(\theta)\sin\alpha_c \qquad (3.2.8)$$

where $\rho(\theta)$ is the curvature radius of the centrode. Figure 3.2.3 illustrates the evolutes σ_2 and σ_2^* of an elliptical centrode at point M.

3.3 Local Representation of a Noncircular Gear

Various teeth of the same noncircular gear have various profiles. This is the main difference between a circular and noncircular gear. The variation of tooth profiles

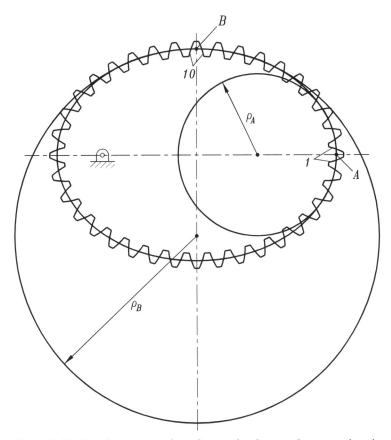

Figure 3.3.1. Local representation of a noncircular gear by respective circular gears.

of the same noncircular gear is the result of variation of curvature of centrode of a noncircular gear.

Visualization of tooth profiles of a noncircular gear is based on the concept that the tooth profiles of the same noncircular gear may be represented as tooth profiles of a set of circular gears. Using such an approach, for instance with an elliptical gear (Fig. 3.3.1), we may represent the profiles of teeth 1 and 10 as profiles of teeth of circular gears with radii ρ_A and ρ_B, respectively. The radii ρ_A and ρ_B are the curvature radii ρ_A and ρ_B at respective points A and B of elliptical centrode for teeth 1 and 10 (Fig. 3.3.1).

The idea of local representation of a noncircular gear is useful for an approximate solution of avoidance of undercutting of noncircular gears (see Section 4.3.2.2).

3.4 Pressure Angle

The conditions of force transmission by noncircular gears are varied in the process of motion due to variation of orientation of profile normal. Analysis of force transmission by noncircular gears require estimation of the range of the *pressure angle*, which is illustrated by drawings of Fig. 3.4.1. The drawings show that (i) the tooth

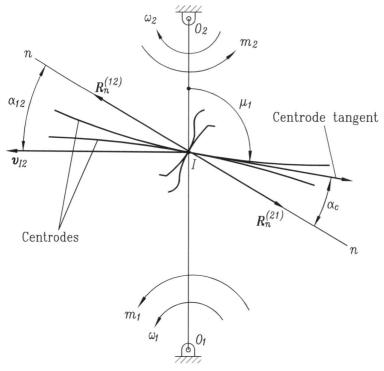

Figure 3.4.1. Illustration of the pressure angle of noncircular gears.

profiles are in contact at point I, the instantaneous center of rotation; (ii) the driving and driven gears are loaded with torques m_1 and m_2, respectively; and (iii) reaction $\mathbf{R}_n^{(12)}$ is transmitted to the driven profile and is directed along the common normal to the tooth profiles (friction of profiles is neglected).

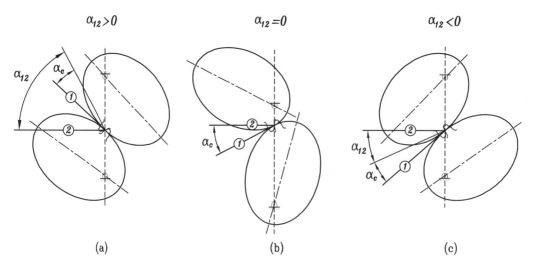

Figure 3.4.2. Variation of the pressure angle α_{12} at position of gears shown in (a), (b), and (c).

The pressure angle α_{12} (Fig. 3.4.1) is formed by $\mathbf{R}_n^{(12)}$ and the velocity $\mathbf{v}_{12}$ of the driven point. It is given by

$$\alpha_{12}(\theta_1) = \mu(\theta_1) + \alpha_c - \frac{\pi}{2} \qquad (3.4.1)$$

wherein the driving profile is the left-side one, as shown in Fig. 3.4.1, and

$$\alpha_{12}(\theta_1) = \mu(\theta_1) - \alpha_c - \frac{\pi}{2} \qquad (3.4.2)$$

wherein the driving profile is the right-side one.

The pressure angle is varied in the process of meshing due to change of angle μ (due to the change of orientation of the tangent to the centrodes). Figure 3.4.2 illustrates the variation of the pressure angle. Figures 3.4.2(a), (b), and (c) show pressure angles $\alpha_{12} > 0, \alpha_{12} = 0, \alpha_{12} < 0$; the tangent to the centrodes and velocity v_{12} are denoted in drawings by 1 and 2, respectively (see graphs of variation of pressure angle α_{12} for elliptical gears in Figs. 4.3.4 and 4.3.5).

4 Elliptical Gears and Gear Drives

4.1 Introduction

The contents of the chapter cover:

(1) The concept of conventional elliptical and modified elliptical centrodes.
(2) Computerized design of gear drives with conventional and modified elliptical centrodes.
(3) Application of theory of planar gearing for advanced design: (i) avoidance of undercutting, and (ii) functional for obtaining identical centrodes.

4.2 Basic Concepts

4.2.1 Ellipse Parameters

Figure 4.2.1 shows (i) the major and minor axes of an ellipse denoted as $2a$ and $2b$, (ii) focuses F and F^* of distance $2c < 2a$, and (iii) chordal distance $2p$. The main feature of the ellipse is that two position vectors $|\overline{FM}| = r_1$ and $|\overline{F^*M}| = r_1^*$ are related as

$$r_1 + r_1^* = 2a \tag{4.2.1}$$

It follows from Eq. (4.2.1) that an ellipse may be traced out by point M of thread $F^* - M - F$ (Fig. 4.2.1) by variation of the location of point M on the ellipse to be generated.

Using Eq. (4.2.1) and drawings of Figs. 4.2.2(a) and 4.2.2(b), it is easy to verify that

$$b = \sqrt{a^2 - c^2} = a\sqrt{1 - e^2} \tag{4.2.2}$$

$$p = a(1 - e^2) \tag{4.2.3}$$

where $e = \dfrac{c}{a}$ is the so-called *ellipse eccentricity*.

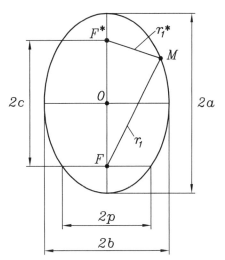

Figure 4.2.1. Generation of ellipse by point M of thread $F^* - M - F$.

4.2.2 Polar Equation of an Ellipse

The polar equation of an ellipse represents ellipse point $M_i (i = 1, 2)$ by the position vector $r_1(\theta_1)$ that is drawn from the respective focus of the ellipse. Henceforth, we will consider two polar equations of an ellipse that correspond to the drawings of Fig. 4.2.3(a), where the polar axis is $\overline{FA_1}$ and the position vector is $\overline{FM_1} = r_1(\theta_1)$, and Fig. 4.2.3(b), where the polar axis is $\overline{F^*A_2}$ and the position vector is $\overline{F^*M_2} = r_1(\theta_1)$. Angle θ_1 in Figs. 4.2.3(a) and (b) is measured clockwise and is formed by the polar axes $\overline{FA_1}$ and $\overline{F^*A_2}$, and position vectors $|\overline{FM_1}|$ and $|\overline{F^*M_2}|$, respectively.

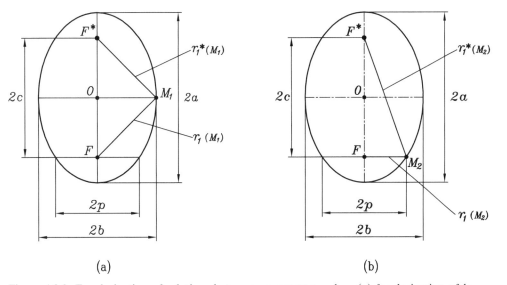

(a)   (b)

Figure 4.2.2. For derivation of relations between parameters a, b, p (a) for derivation of b; (b) for derivation of p.

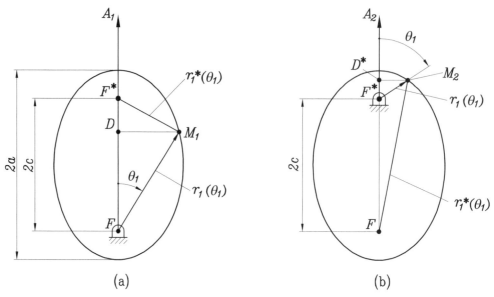

Figure 4.2.3. For derivation of polar equation of an ellipse for the following conditions: (a) the polar axis is $\overline{F A_1}$ and the position vector $\overline{F M_1} = r_1(\theta_1)$; (b) the polar axis is $\overline{F^* A_2}$ and the position vector is $\overline{F^* M_2} = r_1(\theta_1)$.

Drawings of Fig. 4.2.3(a) yield

$$|\overline{F^* D}|^2 + |\overline{D M_1}|^2 = |\overline{F^* M_1}|^2 \tag{4.2.4}$$

Here,

$$|\overline{F^* D}| = 2c - r_1(\theta_1) \cos \theta_1 \tag{4.2.5}$$

$$|\overline{D M_1}| = r_1(\theta_1) \sin \theta_1 \tag{4.2.6}$$

$$|\overline{F^* M_1}| = r_1^*(\theta_1) = 2a - r_1(\theta_1) \tag{4.2.7}$$

as follows from Eq. (4.2.1). Equations from (4.2.4) to (4.2.7), after some transformations, yield

$$r_1(\theta_1) = \frac{a^2 - c^2}{a - c \cos \theta_1} \tag{4.2.8}$$

Taking into account in Eq. (4.2.8) that $e = c/a$ and $p = a(1 - e^2)$, we can obtain the following expression of the polar equation of the ellipse:

$$r_1(\theta_1) = \frac{p}{1 - e \cos \theta_1} \tag{4.2.9}$$

where the polar axis is $\overline{F A_1}$ and the pole is the focus F of the ellipse (Fig. 4.2.3(a)).

Figure 4.2.3(b) illustrates the derivation of the polar equation of the ellipse, wherein the pole is focus F^*. The drawings show that

$$|\overline{F D^*}|^2 + |\overline{D^* M_2}|^2 = |\overline{F M_2}|^2 \tag{4.2.10}$$

Here,

$$|\overline{FD^*}| = 2c + r_1(\theta_1)\cos\theta_1 \qquad (4.2.11)$$

$$|\overline{D^*M_2}| = r_1(\theta_1)\sin\theta_1 \qquad (4.2.12)$$

$$|\overline{FM_2}| = r_1^*(\theta_1) = 2a - r_1(\theta_1) \qquad (4.2.13)$$

Equations from (4.2.10) to (4.2.13), after some transformations, yield the following expression of the polar equation of the ellipse:

$$r_1(\theta_1) = \frac{p}{1 + e\cos\theta_1} \qquad (4.2.14)$$

where the polar axis is $\overline{F^*A_2}$ and the pole is the focus F^* of the ellipse (Fig. 4.2.3(b)).

4.3 External Elliptical Gear Drives

The term "external" means that the driving and driven centrodes, σ_1 and σ_2, are in *external meshing*. Internal gear drives are considered in Chapter 7, wherein the driven centrode σ_2 is in internal meshing with driving centrode σ_1.

The following types of centrodes for external gear drives are considered: (i) conventional elliptical centrodes (Sections 4.3.2 and 4.3.3), (ii) modified elliptical centrodes (Section 4.3.4), (iii) oval centrodes (Section 4.3.5), and (iv) centrodes with lobes (Section 4.3.6).

A conventional elliptical centrode is obtained from a conventional ellipse represented by polar equation (4.2.9), wherein the polar axis is $\overline{FA_1}$ and the pole is the focus F of the ellipse (Fig. 4.2.3(a)), or by polar equation (4.2.14), wherein the polar axis is $\overline{F^*A_2}$ and the pole is the focus F^* of the ellipse (Fig. 4.2.3(b)).

A modified elliptical centrode is obtained from a conventional ellipse, wherein the respective position vectors of the ellipse and modified centrode have the same magnitude but different polar angles (see Section 4.3.4). An oval centrode is obtained from a conventional ellipse as well by the modification described previously (Section 4.3.5). A gear or centrode with lobes is formed by arcs of a modified ellipse (Section 4.3.6).

Design of gear drives with elliptical centrodes (conventional or modified) will cover the case wherein the pinion performs n revolutions for one revolution of the driven gear. The centrodes of the gear drives might be designed with closed or non-closed centrodes.

4.3.1 Basic Equations

The basic equations applied for the design of external elliptical gear drives are as follows:

Step 1. The driving centrode is represented in polar form by

$$r_1(\phi_1), \qquad 0 \le \phi_1 \le 2\pi n \qquad (4.3.1)$$

if the pinion performs n revolutions for one revolution of the driven centrode; ϕ_1 is the angle of rotation of the pinion that is measured opposite to parameter θ_1.

Step 2. The derivative function $m_{21}(\phi_1)$ is represented by

$$m_{21}(\phi_1) = \frac{d\phi_2}{d\phi_1} = \frac{r_1(\phi_1)}{r_2(\phi_1)} = \frac{r_1(\phi_1)}{E - r_1(\phi_1)} \qquad (4.3.2)$$

where $E = r_1(\phi_1) + r_2(\phi_1)$ is the center distance. Similarly, derivative function $m_{12}(\phi_1)$ is represented by

$$m_{12}(\phi_1) = \frac{d\phi_1}{d\phi_2} = \frac{r_2(\phi_1)}{r_1(\phi_1)} = \frac{E - r_1(\phi_1)}{r_1(\phi_1)} \qquad (4.3.3)$$

The meaning of Eqs. (4.3.2) and (4.3.3) is that centrodes σ_1 and σ_2 roll over each other. Function $m_{21}(\phi_1)$ must be a periodic one to obtain a closed form curve (see Section 2.6).

Step 3. Determination of center distance E. It follows from Eqs. (4.3.2) that

$$\phi_2(\phi_1) = \frac{2\pi}{n} = \int_0^{2\pi} m_{21}(\phi_1)d\phi_1 = \int_0^{2\pi} \frac{r_1(\phi_1)}{E - r_1(\phi_1)}d\phi_1 \qquad (4.3.4)$$

where n is an integer number and represents the number of revolutions of the driving centrode for one revolution of the driven centrode. The sought-for center distance is determined as the solution of Eq. (4.3.4) for $E = E(a, e, n)$, where $2a$ is the magnitude of the major axis of the elliptical centrode σ_1 and e is the eccentricity.

It will be proven that

$$E = E(a, e, n) \qquad (4.3.5)$$

is the same for elliptical, modified elliptical, or oval gear drives as well as for gears with lobes. Determination of E for gear drives mentioned here is obtained analytically by application of Eq. (2.6.5) provided by Dwight (Dwight, 1961).

Step 4. Determination of centrode σ_2. Centrode σ_2 is determined by the equations

$$r_2(\phi_1) = E - r_1(\phi_1) \qquad (4.3.6)$$

$$\phi_2(\phi_1) = \int_0^{\phi_1} m_{21}(\phi_1)d\phi_1 \qquad (4.3.7)$$

Equation (4.3.7) represents the transmission function $\phi_2(\phi_1)$ that may be obtained numerically, or analytically by using Eq. (2.6.6) provided by Dwight (Dwight, 1961).

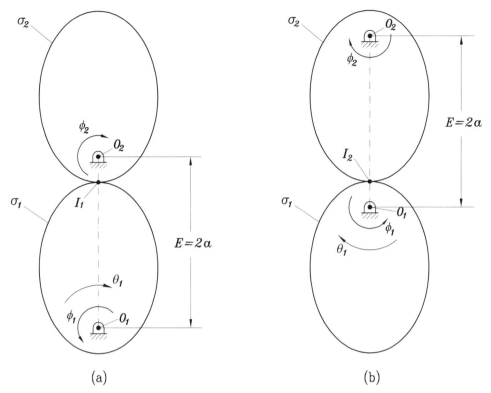

(a)   (b)

Figure 4.3.1. Illustration of two initial positions of gear 1 and 2 where O_1 is (a) the lower focus of ellipse 1; (b) the upper focus of ellipse 1.

4.3.2 Conventional Elliptical Gear Drives

4.3.2.1 Centrodes and Transmission Function

A conventional elliptical gear drive is formed by two elliptical gears that may be assembled as shown in Figs. 4.3.1(a) or 4.3.1(b). Figure 4.3.1(a) shows a gear drive wherein center O_1 of rotation of centrode σ_1 coincides with the lower focus F of ellipse 1. On the other side, Figure 4.3.1(b) shows a gear drive wherein center O_1 of rotation of centrode σ_1 coincides with the upper focus F^* of ellipse 1 (see Section 4.2.2). The polar axes $\overline{FA_1}$ and $\overline{F^*A_2}$ (Fig. 4.2.3) are directed along center distance $\overline{O_1 O_2}$ (Fig. 4.3.1).

Figure 4.3.2 shows an elliptical centrode that performs rotation around center O_1 (it is the lower focus F of ellipse 1). We recall that the instantaneous point I of tangency of centrode 1 and centrode 2 (centrode 2 is not shown) moves along center distance $\overline{O_1 O_2}$ (Fig. 4.3.2) in the process of motion. The center distance is $E = 2a$, where $2a$ is the major axis of mating elliptical centrodes 1 and 2.

Centrode σ_1 (Figs. 4.3.1(a) and 4.3.2) is represented by

$$r_1(\phi_1) = \frac{p}{1 - e\cos\phi_1} = \frac{a(1 - e^2)}{1 - e\cos\phi_1} \qquad (4.3.8)$$

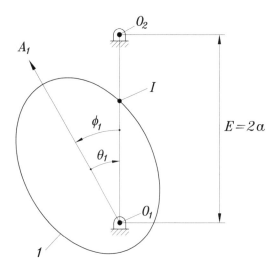

$$E = 2a$$

Figure 4.3.2. For derivation of transmission function $\phi_2(\phi_1)$ of elliptical gears.

The derivative function $m_{12}(\phi_1)$ is determined as

$$m_{12}(\phi_1) = \frac{d\phi_1}{d\phi_2} = \frac{r_2(\phi_1)}{r_1(\phi_1)} = \frac{E - r_1(\phi_1)}{r_1(\phi_1)} \tag{4.3.9}$$

where $r_2(\phi_1)$ determines the magnitude $r_2(\phi_1) = \overline{O_2 I}$ (Fig. 4.3.2). Taking into account Eq. (4.3.8), $E = 2a$, and $p = a(1 - e^2)$, we obtain

$$m_{12}(\phi_1) = \frac{1 + e^2 - 2e \cos\phi_1}{1 - e^2} \tag{4.3.10}$$

The sought-for transmission function may then be represented as

$$\phi_2(\phi_1) = \int_0^{\phi_1} \frac{d\phi_1}{m_{12}(\phi_1)} = (1 - e^2) \int_0^{\phi_1} \frac{d\phi_1}{1 + e^2 - 2e \cos\phi_1} \tag{4.3.11}$$

We obtain transmission function $\phi_2(\phi_1)$ as a closed form solution as follows (Dwight, 1961):

$$\tan\frac{\phi_2}{2} = \frac{1 + e}{1 - e} \tan\frac{\phi_1}{2} \tag{4.3.12}$$

The transmission function represented in Eq. (4.3.12) corresponds to the assembly positions of centrodes as shown in Fig. 4.3.1(a). For assembly as shown in Fig. 4.3.1(b), we obtain

$$\tan\frac{\phi_2}{2} = \frac{1 - e}{1 + e} \tan\frac{\phi_1}{2} \tag{4.3.13}$$

The graph of transmission function $\phi_2(\phi_1)$ given by Eq. (4.3.12) is represented by Fig. 4.3.3. The function is formed by branches (I, II), and (III, IV) that are respectively symmetrical.

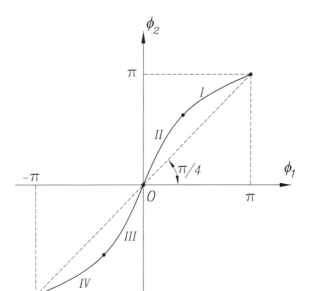

Figure 4.3.3. Transmission function $\phi_2(\phi_1)$ for elliptical gears with $e = 0.4$.

Centrode σ_2 is represented by

$$r_2(\phi_1) = E - r_1(\phi_1) = 2a - \frac{a(1 - e^2)}{1 - e\cos\phi_1} \tag{4.3.14}$$

$$\phi_2(\phi_1) = \int_0^{\phi_1} \frac{d\phi_1}{m_{12}(\phi_1)} = (1 - e^2) \int_0^{\phi_1} \frac{d\phi_1}{1 + e^2 - 2e\cos\phi_1} \tag{4.3.15}$$

4.3.2.2 Influence of Ellipse Parameters and Design Recommendations

VARIATION OF PRESSURE ANGLE α_{12}. The pressure angle α_{12} of noncircular gears has been determined in Section 3.4 by Eqs. (3.4.1) and (3.4.2). It was shown by Fig. 3.4.2 that the sign of α_{12} may be changed in the process of transformation of motions. Figures 4.3.4 and 4.3.5 in addition to Fig. 3.4.2 show the change of magnitude of α_{12} for elliptical gears with various values of eccentricity of the ellipse. It results from investigation of α_{12} that it is necessary to limit the eccentricity of the ellipse to keep α_{12} in the range $-50° \le \alpha_{12} \le 50°$. Function $\alpha_{12max}(e)$ (Fig. 4.3.6) shows the influence of ellipse eccentricity. Reduction of magnitude of α_{12} may be obtained by design of gear drives with more than one stage.

AVOIDANCE OF UNDERCUTTING. The approach is based on the idea of local representation of profiles of noncircular gears by profiles of a set of circular gears (see Section 3.3 and Fig. 3.3.1). The computational procedure is as follows:

(i) We determine the minimal radius ρ_{min} of curvature of an elliptical centrode at point A in Fig. 4.3.7 as

$$\rho_{min} = \frac{b^2}{a} = a(1 - e^2) \tag{4.3.16}$$

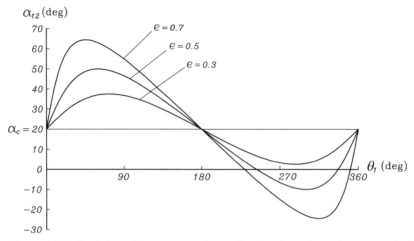

Figure 4.3.4. Variation of pressure angle α_{12} for location of rotation center in lower ellipse focus.

(ii) Undercutting of involute gear of radius $R = \rho_{min}$ is avoided, if the tooth number N is $N > N^*$, where

$$N^* = \frac{2}{\sin^2 \alpha_c} \qquad (4.3.17)$$

where α_c is the profile angle of the rack cutter.

(iii) Consider that the module of the rack cutter is m. Then we obtain the pitch radius of the substituting involute gear as $\rho_A = (mN)/2$. Undercutting is avoided if

$$\frac{mN^*}{2} \leq \rho_{min} = a(1 - e^2) \qquad (4.3.18)$$

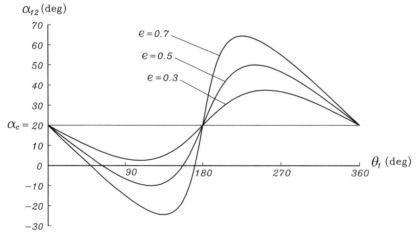

Figure 4.3.5. Variation of pressure angle α_{12} for location of rotation center in upper ellipse focus.

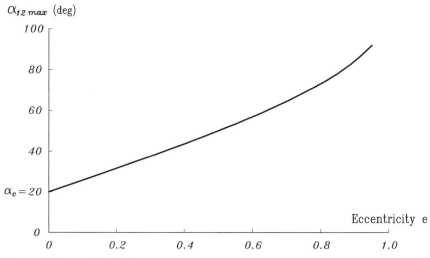

Figure 4.3.6. Maximal value $\alpha_{12_{max}}$ of pressure angle.

This yields the following condition of avoidance of undercutting of elliptical gears as

$$\frac{m}{a} \leq (1 - e^2) \sin^2 \alpha_c \qquad (4.3.19)$$

The developed approach may be applied for other types of noncircular gears, but it needs determination of the minimal curvature radius of the centrode of the gear.

APPLICATION OF ODD TOOTH NUMBER OF ELLIPTICAL GEARS. While choosing the linear dimensions of elliptical gears, it is required that the length L of the centrode of an elliptical gear will be equal to

$$L = \pi m N \qquad (4.3.20)$$

where m is the module of the tool and N the number of teeth of the elliptical gear.

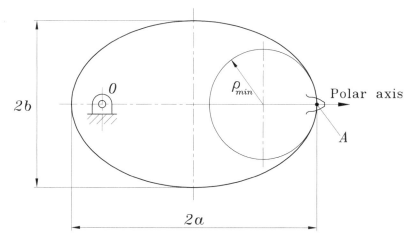

Figure 4.3.7. For determination of minimal curvature radius ρ_{min} of elliptical centrode.

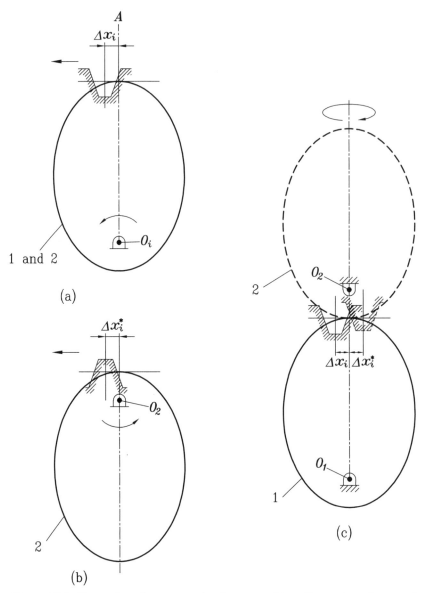

Figure 4.3.8. Illustration of compensation for error of installment Δx_i of generating tool.

Figure 4.3.8(a) illustrates that the axis of symmetry of the tooth of the generating rack cutter is dislocated on distance Δx_i ($i = 1, 2$) due to an error of installment, while gears 1 and 2 are being generated. Compensation of the error Δx_i of the installment of the tool requires (i) to choose an odd number of teeth for both elliptical gears 1 and 2, and (ii) make a turn of $180°$ about $\overline{O_1 O_2}$ of one of the gears in the assembly. This statement is based on the following considerations:

(i) It is assumed that the same hob (rack cutter) is applied for generation of the mating elliptical gears and they have the same number of teeth.

(ii) Figure 4.3.8(a) shows the error Δx_i of installment of the rack cutter with respect to line $\overline{O_i A}$. The center distance of the gears after their assembly will coincide with $\overline{O_i A}$.

(iii) Figure 4.3.8(b) shows the location of the tooth of the rack cutter after its displacement a distance

$$L^* = \frac{\pi m N}{2} \tag{4.3.21}$$

where N is the odd number of teeth of the elliptical gear. Here L^* is measured from the initial position of the rack cutter shown in Fig. 4.3.8(a). We recall that Δx_i denotes the displacement of the axis of symmetry of the space of the rack cutter (Fig. 4.3.8(a)) and Δx_i^* denotes the displacement of the tooth of the rack cutter (Fig. 4.3.8(b)) after a rack cutter translation of magnitude L^*.

(iv) Figure 4.3.8(c) shows that the axes of symmetry of the respective space and tooth of the rack cutter do not coincide, although $|\Delta x_i| = |\Delta x_i^*|$. The compensation of the error of installment of the rack cutter will be achieved after one of the mating gears is turned on $180°$ around $\overline{O_1 O_2}$, as shown in Fig. 4.3.8(c).

TOOTH NUMBER OF ELLIPTICAL GEARS. The number of teeth that might be provided for an elliptical gear is determined by the parameters of the ellipse (dimensions of major and minor axes a and b) and the module m. The first step of the computational procedure is the determination of the length of the elliptical centrode considering the ellipse as the pitch curve.

To determine the length of the elliptical centrode, it is preferable to represent the centrode in parametric form as

$$x = a \sin \psi, \qquad y = b \cos \psi \tag{4.3.22}$$

The arc ds of the ellipse is represented as

$$ds = \sqrt{dx^2 + dy^2} = \sqrt{a^2 \cos^2 \psi + b^2 \sin^2 \psi}\, d\psi = a\sqrt{1 - e^2 \sin^2 \psi}\, d\psi \tag{4.3.23}$$

The length of an arc of the centrode is determined by

$$s = a \int_0^\psi \sqrt{1 - e^2 \sin^2 \psi}\, d\psi, \qquad e = \frac{\sqrt{a^2 - b^2}}{a} = \frac{c}{a} \tag{4.3.24}$$

The integral

$$\int_0^\psi \sqrt{1 - e^2 \sin^2 \psi}\, d\psi \tag{4.3.25}$$

has found a broad application in astronomy as the elliptic integral of second kind and may be calculated numerically.

The number of teeth of an elliptical gear is determined by

$$N = \frac{4a}{m\pi} \int_0^{\frac{\pi}{2}} \sqrt{1 - e^2 \sin^2 \psi}\, d\psi \tag{4.3.26}$$

where $m\pi$ is the distance between the neighboring teeth measured along the centrode. Considering that module m is of a standard value, we may obtain an integer number N of teeth by taking a proper magnitude for a. Considering that the elliptical centrode is represented by

$$r(\theta) = \frac{a(1 - e^2)}{1 + e\cos\theta} \tag{4.3.27}$$

we obtain the following relation between parameters ψ and θ:

$$\cos\theta = \frac{\sin\psi - e}{1 - e\sin\psi} \tag{4.3.28}$$

4.3.3 Gear Drive with Elliptical Pinion and Conjugated Gear

The difference with the gear drive of previous section is that pinion σ_1 performs not one but n revolutions for one revolution of gear σ_2. To obtain σ_2 as a closed-form curve, the center distance E must satisfy the equation derived in Section 2.6.

The computational procedure is based on the following equations:

(1) Centrode σ_1 is determined as

$$r_1(\phi_1) = \frac{a(1 - e^2)}{1 - e\cos\phi_1} \tag{4.3.29}$$

(2) Center distance E is determined as (see Problem 2.6.1 in Section 2.6)

$$E = a\left[1 + \sqrt{1 + (n^2 - 1)(1 - e^2)}\right] \tag{4.3.30}$$

It may be noticed that for $n = 1$, $E = 2a$, as considered for conventional elliptical gear drives.

(3) The derivative function is

$$m_{21}(\phi_1) = \frac{r_1(\phi_1)}{E - r_1(\phi_1)} = \left(\frac{1}{n}\right)\frac{1 - e^2}{1 - 2e\cos\phi_1 + e^2} \tag{4.3.31}$$

(4) Centrode σ_2 is represented by equations

$$r_2(\phi_1) = E - r_1(\phi_1) \tag{4.3.32}$$

$$\phi_2(\phi_1) = \int_0^{\phi_1} m_{21}(\phi_1)d\phi_1 \tag{4.3.33}$$

where $\phi_2(\phi_1)$ is the transmission function. After derivations, we obtain

$$\tan\frac{n\phi_2}{2} = \frac{1 + e}{1 - e}\tan\frac{\phi_1}{2} \tag{4.3.34}$$

We emphasize that the centrode of gear 1 is a conventional ellipse, but the centrode of gear 2 for $n \neq 1$ is a transformed ellipse.

Figures 4.3.9(a) and 4.3.9(b) illustrate the centrodes of two gear drives in which centrode σ_1 is represented as a conventional ellipse that performs two and three revolutions, respectively, for one revolution of centrode σ_2.

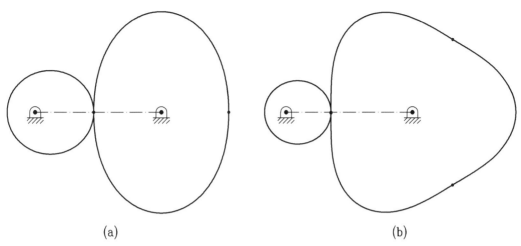

(a)   (b)

Figure 4.3.9. Illustration of conjugation of a conventional ellipse ($e = 0.35$) with the conjugated transformed ellipse for (a) $n = 2$ and (b) $n = 3$.

4.3.4 External Gear Drive with Modified Elliptical Gears

Modification of the centrodes enables obtaining an asymmetric derivative function $m_{21}(\phi_1) = \omega^{(2)}/\omega^{(1)}$ of the gear ratio of the gear drive (Fig. 4.3.10). The interval I: $0 \le \phi_1 \le \pi/m_I$ ($\pi/m_I < \pi$) corresponds to the free-running part of the cycle (faster), and the interval II: $\pi/m_I \le \phi_1 \le 2\pi$ corresponds to the working part of the cycle (slower). Application of a gear drive with gears having such centrodes is useful for tandem design with a linkage (see Section 10.6.2) for variation of the output speed.

4.3.4.1 Modification of the Ellipse

The modified elliptical centrode of the driving gear 1 is formed by two branches, I and II (Fig. 4.3.11). The transformation of the conventional elliptical centrode into two modified branches is performed with observation of the following relations:

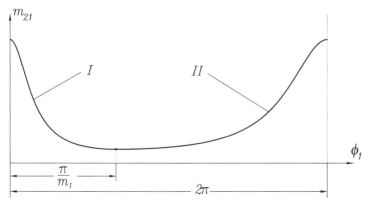

Figure 4.3.10. Asymmetric function $m_{21}(\phi_1)$ of gear ratio of modified elliptical gears.

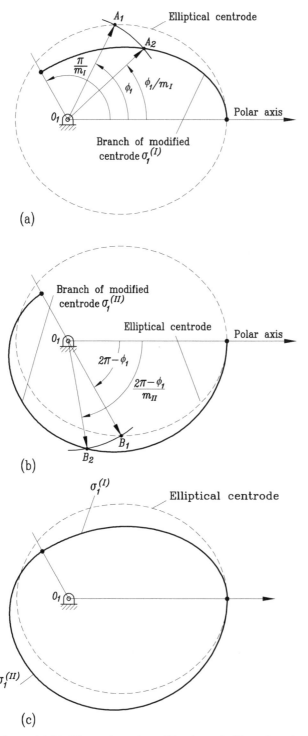

Figure 4.3.11. Illustration of modification of ellipse for obtaining the centrode of gear 1:
(a) change of polar angle from ϕ_1 to $\frac{\phi_1}{m_I}$; (b) change of polar angle from $(2\pi - \phi_1)$ to $(\frac{2\pi - \phi_1}{m_{II}})$;
(c) illustration of the two branches $\sigma_1^{(I)}$ and $\sigma_1^{(II)}$ of the modified centrode.

(i) The initial and modified vector of the upper branch (Fig. 4.3.11(a)) are $\overline{O_1 A_1}$ and $\overline{O_1 A_2}$, wherein $|\overline{O_1 A_2}| = |\overline{O_1 A_1}|$, but the polar angle is changed from ϕ_1 to ϕ_1/m_I, where m_I is the coefficient of modification of the branch I of the ellipse.

(ii) Similarly, for the other branch of the centrode (Fig. 4.3.11(b)) we have $|\overline{O_1 B_2}| = |\overline{O_1 B_1}|$ but the polar angle for position vector $\overline{O_1 B_2}$ is changed from $2\pi - \phi_1$ to $(2\pi - \phi_1)/m_{II}$. The coefficients of modification m_I and m_{II} are related by

$$m_{II} = \frac{m_I}{2m_I - 1} \tag{4.3.35}$$

This yields that the two branches (Figs. 4.3.11(a) and 4.3.11(b)) have a common point at $\phi_1 = \pi/m_I$.

Modified centrode σ_1 is formed by curves $\sigma_1^{(I)}$ and $\sigma_1^{(II)}$ (Fig. 4.3.11(c)) that are in tangency at their common point.

4.3.4.2 Derivation of Modified Centrode σ_1

The polar equation of a conventional ellipse is represented as

$$r_1(\phi_1) = \frac{p}{1 - e\cos\phi_1}, \qquad 0 \le \phi_1 \le 2\pi \tag{4.3.36}$$

The two branches of the modified ellipse are represented with

$$r_1^{(I)}(\phi_1) = \frac{p}{1 - e\cos(m_I\phi_1)}, \qquad 0 \le \phi_1 \le \frac{\pi}{m_I} \tag{4.3.37}$$

$$r_1^{(II)}(\phi_1) = \frac{p}{1 - e\cos(m_{II}(2\pi - \phi_1))}, \qquad \frac{\pi}{m_I} \le \phi_1 \le 2\pi \tag{4.3.38}$$

Here, $p = a(1 - e^2)$.

4.3.4.3 Derivative Functions $m_{21}^{(I)}(\phi_1)$ and $m_{21}^{(II)}(\phi_1)$

The derivation is based on the following relations:

$$m_{21}(\phi_1) = \frac{\omega^{(2)}}{\omega^{(1)}} = \frac{r_1(\phi_1)}{r_2(\phi_1)} = \frac{r_1(\phi_1)}{E - r_1(\phi_1)} \tag{4.3.39}$$

We then obtain

$$m_{21}^{(I)}(\phi_1) = \frac{p}{E - p - Ee\cos(m_I\phi_1)}, \qquad 0 \le \phi_1 \le \frac{\pi}{m_I} \tag{4.3.40}$$

$$m_{21}^{(II)}(\phi_1) = \frac{p}{E - p - Ee\cos(m_{II}(2\pi - \phi_1))}, \qquad \frac{\pi}{m_I} \le \phi_1 \le 2\pi \tag{4.3.41}$$

4.3.4.4 Relation between Rotations of Gears 1 and 2

Rotations of gear 1 and 2 are related by

$$2\pi = n\left[\int_0^{\pi/m_I} m_{21}^{(I)}(\phi_1)d\phi_1 + \int_{\pi/m_I}^{2\pi} m_{21}^{(II)}(\phi_1)d\phi_1\right] \tag{4.3.42}$$

The meaning of Eq. (4.3.42) is that pinion 1 (centrode σ_1) will perform n revolutions for one revolution of gear 2 (centrode σ_2).

The following derivations are based on the change of variables:

$$m_I \phi_1 = x, \qquad d\phi_1 = \frac{dx}{m_I} \tag{4.3.43}$$

$$m_{II}(2\pi - \phi_1) = y, \qquad d\phi_1 = -\frac{dy}{m_{II}} \tag{4.3.44}$$

Then, Eq. (4.3.42) may be represented as

$$\frac{2\pi}{n} = \left(\frac{p}{E-p}\right)\left[\left(\frac{1}{m_I}\right)\int_0^\pi \frac{dx}{1 - q\cos x} + \left(\frac{1}{m_{II}}\right)\int_0^\pi \frac{dy}{1 - \mu\cos y}\right] \tag{4.3.45}$$

where $q = \mu = \dfrac{Ee}{E-p}$.

From the tables of integrals provided by Dwight (Dwight, 1961) (see Eq. (2.6.5)), we obtain

$$\int_0^\pi \frac{dx}{1 - q\cos x} = \frac{\pi}{\sqrt{1 - q^2}}, \qquad \int_0^\pi \frac{dy}{1 - \mu\cos y} = \frac{\pi}{\sqrt{1 - \mu^2}} \tag{4.3.46}$$

It follows as well from (4.3.35) that

$$\frac{1}{m_I} + \frac{1}{m_{II}} = 2 \tag{4.3.47}$$

The solution of Eq. (4.3.45) for $E = E(a, e, n)$ yields the same equation for the center distance as the one obtained for a gear drive with conventional elliptical gears:

$$E = a[1 + \sqrt{1 + (n^2 - 1)(1 - e^2)}] \tag{4.3.48}$$

4.3.4.5 Derivation of Centrode σ_2

The branches of centrode σ_2 are represented by

$$r_2^{(I)}(\phi_1) = E - r_1^{(I)}(\phi_1)$$

$$\phi_2^{(I)}(\phi_1) = \int_0^{\phi_1} m_{21}^{(I)}(\phi_1)d\phi_1, \qquad 0 \leq \phi_1 \leq \frac{\pi}{m_I} \tag{4.3.49}$$

$$r_2^{(II)}(\phi_1) = E - r_1^{(II)}(\phi_1)$$

$$\phi_2^{(II)}(\phi_1) = \int_0^{\phi_1} m_{21}^{(II)}(\phi_1)d\phi_1, \qquad \frac{\pi}{m_I} \leq \phi_1 \leq 2\pi \tag{4.3.50}$$

The transmission functions $\phi_2^{(i)}(\phi_1)$, $(i = I, II)$, may be obtained by numerical integration or analytically by using Eq. (2.6.6) (Dwight, 1961) for a function

$$f(x) = \frac{1}{a \pm b\cos x}$$

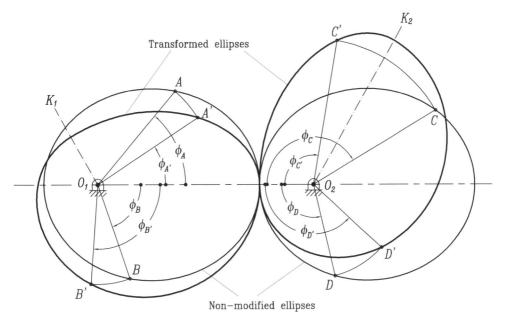

Figure 4.3.12. Illustration of original and modified ellipses and transformation of point A, B, C, and D of original ellipses into points A', B', C', and D' of modified ellipses.

Drawings of Fig. 4.3.12 show the centrodes obtained for parameter $m_I = 3/2$, $n = 1$, $e = 0.5$. The drawings of Fig. 4.3.12 illustrate the transformation of points A, B, C, D of original ellipses to points A', B', C', D' of modified ellipses.

Drawings of Fig. 4.3.13 show the centrodes obtained for parameter $m_I = 3/2$, $e = 2/\sqrt{67}$ for the following cases of design: (a) $n = 1$, (b) $n = 2$, and (c) $n = 3$. Drawings of Fig. 4.3.14 show the centrodes obtained for the same cases of design, wherein the ellipse eccentricity $e = 0.5$ and $m_I = 3/2$.

4.3.5 External Gear Drive with Oval Centrodes

4.3.5.1 Equation of Oval Centrode

An oval centrode is obtained by transformation of a conventional ellipse (see Section 4.3.4) based on application of a parameter of modification m_I as an integer $m_I = 2$. Such oval gears with $m_I = 2$ and $n = 1$ have found application in the past in flow meters for measuring the discharge of liquids (Fig. 4.3.15).

The centrode of an oval gear is represented by

$$r_1(\phi_1) = \frac{a(1 - e^2)}{1 - e \cos 2\phi_1} = \frac{p}{1 - e \cos 2\phi_1} \qquad (4.3.51)$$

The driven centrode 2 is represented by

$$r_2(\phi_1) = E - r_1(\phi_1) = E - \frac{p}{1 - e \cos 2\phi_1} \qquad (4.3.52)$$

We will consider gear drives wherein gear 1 performs one or n revolutions for one revolution of the driven gear.

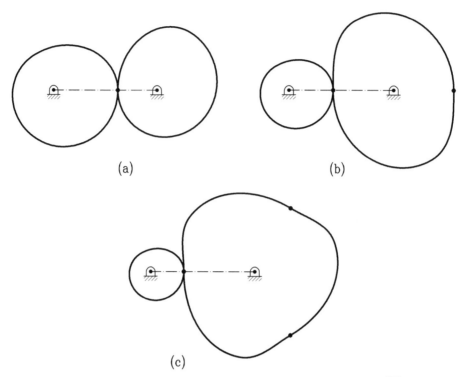

Figure 4.3.13. Centrodes of gears with modified ellipses: $m_I = 3/2$, $e = 2/\sqrt{67}$ and (a) $n = 1$, (b) $n = 2$, and (c) $n = 3$.

4.3.5.2 Derivative Function $m_{21}(\phi_1)$

Following the procedure discussed in Section 4.3.4.3, we determine

$$m_{21}(\phi_1) = \frac{d\phi_2}{d\phi_1} = \frac{r_1(\phi_1)}{E - r_1(\phi_1)} \tag{4.3.53}$$

as

$$m_{21}(\phi_1) = \frac{p}{E - p - Ee \cos 2\phi_1} \tag{4.3.54}$$

4.3.5.3 Relation between Rotations of Gears 1 and 2

This relation is represented as

$$2\pi = n \int_0^{2\pi} \frac{p}{E - p - Ee \cos 2\phi_1} d\phi_1 = \left(\frac{np}{E - p}\right) \int_0^{2\pi} \frac{d\phi_1}{1 - \left(\dfrac{Ee}{E - p}\right) \cos 2\phi_1} \tag{4.3.55}$$

The variable ϕ_1 is changed for x wherein

$$2\phi_1 = x, \qquad d\phi_1 = \frac{dx}{2} \tag{4.3.56}$$

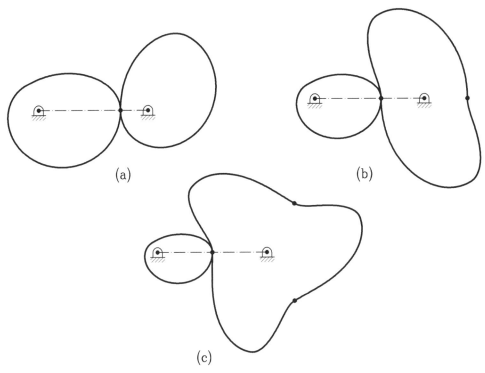

Figure 4.3.14. Centrodes of gears with modified ellipses: $m_I = 3/2$, $e = 0.5$ and (a) $n = 1$, (b) $n = 2$, and (c) $n = 3$.

which yields

$$2\pi = \left(\frac{np}{E-p}\right)\left(\frac{4}{2}\right)\int_0^\pi \frac{dx}{1 - q\cos x} \tag{4.3.57}$$

where $q = \dfrac{Ee}{E-p}$. By applying Eq. (2.6.5) provided by Dwight (Dwight, 1961) and after simple derivations, we obtain equation $E = E(a, e, n)$ as

$$E = a[1 + \sqrt{1 + (n^2 - 1)(1 - e^2)}] \tag{4.3.58}$$

which is the same as Eq. (4.3.48).

Figure 4.3.15. Oval gears in a flow meter.

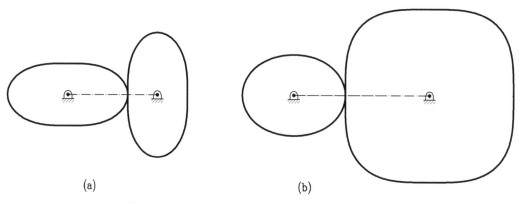

Figure 4.3.16. Illustration of gear drives with oval gears: (a) two oval gears wherein $e = 1/3$ and $n = 1$; (b) oval gear with conjugated modified oval driven gear wherein $e = 1/\sqrt{57}$ and $n = 2$.

4.3.5.4 Transmission Function $\phi_2(\phi_1)$

Transmission function $\phi_2(\phi_1)$ is represented as

$$\phi_2(\phi_1) = \int_0^{\phi_1} \frac{p}{E - p - Ee\cos 2\phi_1} d\phi_1 \qquad (4.3.59)$$

We can consider Eq. (4.3.59) in the following form:

$$\int_0^{\phi_1} \frac{p}{E - p - Ee\cos 2\phi_1} d\phi_1 = \int_0^x \frac{dx}{b + c\cos ax} \qquad (4.3.60)$$

where $a = 2$, $b = E - p$, and $c = -Ee$. By applying Eq. (2.6.6) provided by Dwight (Dwight, 1961), we obtain that Eq. (4.3.60) for $n = 1$ yields

$$\tan \phi_2 = \frac{1 + e}{1 - e} \tan \phi_1 \qquad (4.3.61)$$

For the case when $n \geq 2$, Eq. (4.3.61) is generalized as

$$\tan \phi_2 = \frac{\sqrt{1 + (n^2 - 1)(1 - e^2)} + e}{n(1 - e)} \tan \phi_1 \qquad (4.3.62)$$

Centrodes of oval gears shown in Fig. 4.3.16(a) are designed for the ratio of revolutions $n = 1$ and eccentricity of the original ellipse $e = 1/3$, which allows for providing convexity of the centrode. The centrodes of Fig. 4.3.16(b) correspond to the case when $e = 1/\sqrt{57}$ for avoidance of concavity of centrode 2 and $n = 2$ (revolutions of centrode 1 for one revolution of centrode 2).

4.3.6 Design of Noncircular Gears with Lobes

The term *lobe* used in technical literature has been adapted from anatomy to emphasize that the noncircular gear has several parts. An example of a noncircular gear that has three lobes is shown in Fig. 4.3.17.

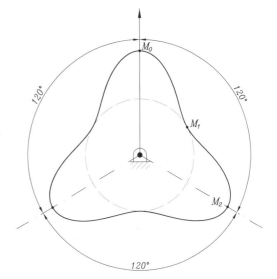

Figure 4.3.17. Illustration of a noncircular gear with three lobes.

From such a point of view, an oval gear may be also considered as a noncircular gear with two lobes. The noncircular gears with lobes that are considered in this section have three or more lobes. We consider the profiles of lobes obtained by modification of ellipses (see ellipse modification in Section 4.3.4).

The driving and driven pinion and gear designed by ellipse modification are in a mixed external-internal meshing. There is a possibility to generate the lobes as gears by application of a shaper with a small number of teeth as done for internal noncircular gears.

This section covers design of noncircular gears with lobes based on the application of conjugated centrodes. Using derivations similar to those applied for oval centrodes, we obtain the following generalized system of equations for the design of noncircular gears with lobes:

$$r_1(\phi_1) = \frac{p}{1 - e \cos m_1 \phi_1} = \frac{a(1 - e^2)}{1 - e \cos m_1 \phi_1} \qquad (4.3.63)$$

$$E = a \left[1 + \sqrt{1 + (n^2 - 1)(1 - e^2)} \right] \qquad (4.3.64)$$

$$r_2(\phi_1) = E - r_1(\phi_1) \qquad (4.3.65)$$

$$\tan \frac{m_2 \phi_2}{2} = \frac{\sqrt{1 + (n^2 - 1)(1 - e^2)} + e}{n(1 - e)} \tan \frac{m_1 \phi_1}{2} \qquad (4.3.66)$$

Here, m_1 and m_2 are the number of lobes of gear 1 and gear 2, respectively, and $n = \dfrac{m_2}{m_1}$ is the ratio of related revolutions of gears 1 and 2.

Figure 4.3.18 shows examples of centrodes of noncircular gears with lobes for (a) $n = 1$ and $m_1 = \{1, 2, 3, 4\}$; (b) $n = 2$ and $m_1 = \{1, 2, 3\}$; (c) $n = 3$ and $m_1 = \{1, 2\}$. For all cases represented, the ellipse eccentricity $e = 0.6$.

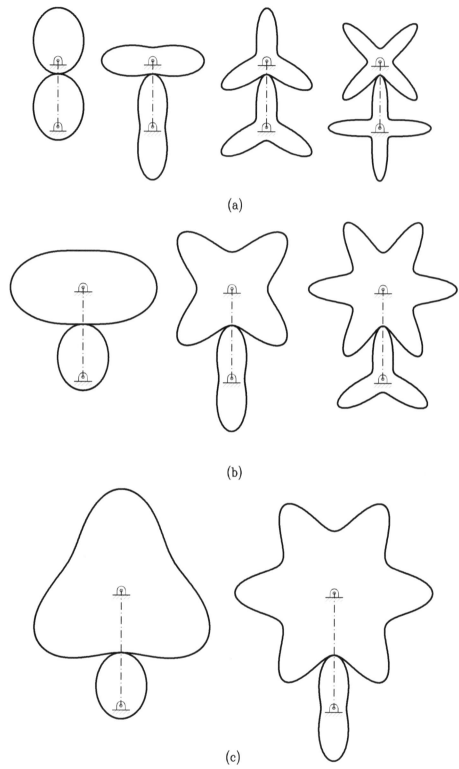

Figure 4.3.18. Examples of centrodes of noncircular gears with lobes: (a) $n = 1$ and $m_1 = \{1, 2, 3, 4\}$; (b) $n = 2$ and $m_1 = \{1, 2, 3\}$; (c) $n = 3$ and $m_1 = \{1, 2\}$.

4.3.6.1 Design of Gear Drives with Different Number of Lobes for Pinion and Gear

Let us consider that the pinion is formed by three lobes and the gear is formed by four lobes. The procedure of design of such gear drive is as follows:

Step 1. Polar Equation of Pinion 1.

The equation is based on transformation of a conventional ellipse (see Section 4.3.2) and is represented as

$$r_1(\phi_1) = \frac{p}{1 - e \cos 3\phi_1}, \qquad p = a(1 - e^2), \qquad 0 \le \phi_1 \le 2\pi \qquad (4.3.67)$$

In comparison with an oval gear, the coefficient of modification is 3 but not 2. The coefficient of modification affects the magnitude of the polar angle of the radius vector, and therefore the shape of the lobe differs substantially from the shape of the ellipse and oval gear. The profiles of all three lobes of the pinion are identical.

The following derivations are directed toward determination of the center distance E between the pinion 1 and gear 2, and derivation of the equation of the gear centrode that is formed by four lobes. The applied procedure is represented as a sequence of steps.

Step 2. Determination of Derivative Function $m_{21}(\phi_1)$.

Function $m_{21}(\phi_1)$ is determined by application of

$$m_{21}(\phi_1) = \frac{d\phi_2}{d\phi_1} = \frac{r_1(\phi_1)}{r_2(\phi_2)} = \frac{r_1(\phi_1)}{E - r_1(\phi_1)} \qquad (4.3.68)$$

where $r_2(\phi_1)$ represents the radius vector of the gear centrode. Observation of Eq. (4.3.68) guarantees that the pinion-gear centrodes roll over each other because the relative velocity is equal to zero.

Equations (4.3.67) and (4.3.68) yield the following expression of $m_{21}(\phi_1)$:

$$m_{21}(\phi_1) = \frac{p}{E - p - Ee \cos 3\phi_1} = \left(\frac{p}{E - p} \right) \left(\frac{1}{1 - q \cos 3\phi_1} \right), \qquad q = \frac{Ee}{E - p}$$

$$(4.3.69)$$

Step 3. Determination of Relation between Revolutions of Pinion and Gear.

We consider in our example that the pinion and gear are provided with three and four lobes, respectively, and therefore the pinion performs four revolutions for three revolutions of the gear. This yields

$$2\pi = n \int_0^{2\pi} m_{21}(\phi_1) d\phi_1 = n \left(\frac{p}{E - p} \right) \left(\frac{1}{3} \right) \int_0^{2\pi} \frac{d(3\phi_1)}{1 - q \cos(3\phi_1)}, \qquad n = \frac{4}{3}$$

$$(4.3.70)$$

The following derivations are directed towards determination of the definite integral (4.3.70) by using the tables of integrals by Dwight (Dwight, 1961). The

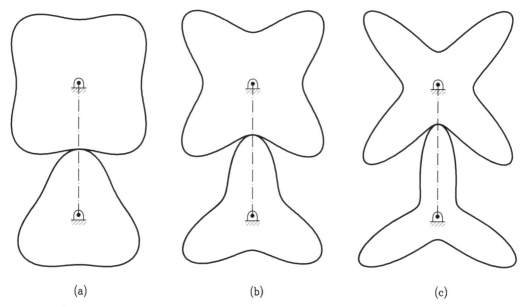

(a)   (b)   (c)

Figure 4.3.19. Illustration of gear drive with three-lobe pinion and four-lobe gear for the
following values of eccentricity of the ellipse: (a) $e = 0.2$, (b) $e = 0.4$, and (c) $e = 0.6$.

derivations are based on a change of variable by using

$$3\phi_1 = x, \qquad dx = \frac{d\phi_1}{3} \tag{4.3.71}$$

and the result of the following integral provided by Dwight (Dwight, 1961):

$$\int_0^\pi \frac{dx}{1 - q\cos x} = \frac{\pi}{\sqrt{1 - q^2}} \tag{4.3.72}$$

Finally, we obtain the function

$$E = E(a, e, n), \qquad n = \frac{4}{3} \tag{4.3.73}$$

as

$$E = a[1 + \sqrt{1 + (n^2 - 1)(1 - e^2)}], \qquad n = \frac{4}{3} \tag{4.3.74}$$

Equation (4.3.74) represents an extension of the previous formulation for determination of E for the case when n is not an integer number but a ratio of
gear-pinion lobes that differs from an integer number.

Step 4. Derivation of Centrode of Driven Gear. Driven gear 2 is represented by
four lobes, and each lobe is determined by

$$r_2(\phi_1) = E - r_1(\phi_1) = E - \frac{p}{1 - e\cos(m_1\phi_1)}, \qquad (m_1 = 3) \tag{4.3.75}$$

$$\phi_2(\phi_1) = n\left(\frac{p}{E - p}\right)\frac{1}{3}\int_0^{\phi_1}\frac{d(m_1\phi_1)}{1 - q\cos(m_1\phi_1)}, \qquad (m_1 = 3) \tag{4.3.76}$$

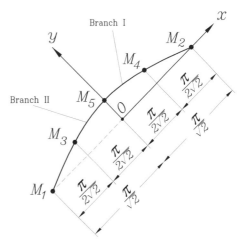

Figure 4.4.1. Set of points M_i ($i = 1, \cdots, 5$) chosen for determination of function $\phi_2(\phi_1)$ as the curve of second order.

Transmission function $\phi_2(\phi_1)$ may be determined analytically, using Eq. (2.6.6) provided by Dwight (Dwight, 1961), or numerically, by integration of Eq. (4.3.76).

Figure 4.3.19 represents several examples of gear drives for different values of the eccentricity wherein the pinion has three lobes and the gear has four lobes.

4.4 Transmission Function of Elliptical Gears as Curve of Second Order

Transmission function $\phi_2(\phi_1)$ of elliptical gears has been represented in closed form by Eq. (4.3.12) (see as well Eq. (4.3.13)). Our goal is to prove that $\phi_2(\phi_1)$ for a gear drive with identical ellipses is a curve of second order, particularly, a hyperbola. The proof is based on the following procedure:

Step 1. It is known from analytical geometry that curve σ of second order may be represented in coordinate system $S(x, y)$ by an equation of the following structure (Korn & Korn, 1968):

$$a_{11}x^2 + 2a_{12}xy + a_{22}y^2 + 2a_{13}x + 2a_{23}y + a_{33} = 0 \qquad (4.4.1)$$

Curve σ may be determined by choosing five points, but any four of these five points should not belong to a straight line.

Step 2. The procedure is applied for transmission function $\phi_2(\phi_1)$ determined by Eq. (4.3.12). The chosen five points belong to branches I and II and are represented by Fig. 4.4.1 as the set of points M_i ($i = 1, \cdots, 5$). Thus,

$$M_1\left(-\tfrac{\pi}{\sqrt{2}}, 0\right), \quad M_2\left(\tfrac{\pi}{\sqrt{2}}, 0\right), \quad M_3\left(-\tfrac{\pi}{2\sqrt{2}}, y_3\right), \quad M_4\left(\tfrac{\pi}{2\sqrt{2}}, y_4\right), \quad M_5\left(0, y_5\right)$$
$$(4.4.2)$$

Here, $y_4 = y_3$ because branches I and II are symmetrical.

Points M_i ($i = 1, \cdots, 5$) must belong to the sought-for curve σ (of second order), and this is stated as observation of a system of five linear equations.

The derivation of five linear equations is based on Eq. (4.4.1). Thus, we have

$$a_{11}x_i^2 + 2a_{12}x_i y_i + a_{22}y_i^2 + 2a_{13}x_i + 2a_{23}y_i + a_{33} = 0 \quad (i = 1, \cdots, 5) \quad (4.4.3)$$

The unknowns in Eq. (4.4.3) are the coefficients $a_{k\ell}$ ($k = 1, 2, 3; \ell = 1, 2, 3$).

Step 3. The solutions of Eq. (4.4.3) for $a_{k\ell}$ yield

$$a_{11} = -\frac{2}{\pi^2} a_{33},$$

$$a_{22} = \frac{\frac{3}{2}y_5 - 2y_3}{2y_3 y_5(y_5 - y_3)} a_{33},$$

$$a_{23} = -\frac{\frac{3}{2}y_5^2 - 2y_3^2}{4y_3 y_5(y_5 - y_3)} a_{33},$$

$$a_{12} = a_{13} = 0$$

(4.4.4)

Step 4. We consider now that a point with coordinates (x, y) belongs to the sought-for curve σ of second order. We may then represent σ by using Eq. (4.4.1) and coefficients (4.4.4) as follows

$$\frac{2}{\pi^2}x^2 - \frac{\frac{3}{2}y_5 - 2y_3}{2y_3 y_5(y_5 - y_3)}y^2 + \frac{\frac{3}{2}y_5^2 - 2y_3^2}{2y_3 y_5(y_5 - y_3)}y = 1 \quad (4.4.5)$$

Observation of Eq. (4.4.5) is the proof that the transmission function $\phi_2(\phi_1)$ is indeed a curve of second order. We recall that $\phi_2(\phi_1)$ has been previously determined by nonlinear Eq. (4.3.13).

Step 5. In addition, it is necessary to determine the type of the second-order curve σ that might be an ellipse, a hyperbola, or a parabola. The solution is based on determination of the invariants of Eq. (4.4.4) (Korn & Korn, 1968). The investigation shows that

$$D = \begin{vmatrix} a_{11} & a_{12} \\ a_{12} & a_{22} \end{vmatrix} < 0 \quad (4.4.6)$$

and therefore σ (represented by Eq. (4.4.5)) is a hyperbola. This conclusion has been confirmed as well numerically for elliptical gear drives with various values of ellipse eccentricity of the identical centrodes.

4.5 Functional of Identical Centrodes

Some noncircular gear drives have identical centrodes – for instance, elliptical gear drives, oval gear drives, gear drives for generation of function $1/x$, or those proposed by Tong and Yang (Tong & Yang, 1998).

Transmission function of gear drives with identical centrodes has a specific structure that may be recognized considering, for instance, function $\phi_2(\phi_1)$ for elliptical gear drives (Fig. 4.3.3). In this case, function $\phi_2(\phi_1)$ is formed by symmetry of branches (I, II) and (III, IV), respectively. Figures 4.5.1(a) and (b) illustrate

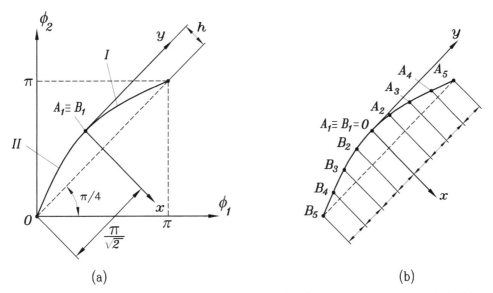

Figure 4.5.1. Illustration of symmetry of function $\phi_2(\phi_1)$: (a) branches I and II of $\phi_2(\phi_1)$; (b) representation of $\phi_2(\phi_1)$ in coordinate system (x, y).

the symmetry of branches I and II. Similar illustrations may be provided as well for branches (III, IV) (see Fig. 4.3.3).

The analysis of function $\phi_2(\phi_1)$ for elliptical gears is based on the following considerations:

(1) Figure 4.5.2 represents two small pieces of transmission function $\phi_2(\phi_1)$ with the origin $O \equiv B_5$ (see Fig. 4.5.1(b)) and function $\Phi_1(\Phi_2)$ with the origin $O' \equiv A_5$ (see Fig. 4.5.1 (b)). If identical centrodes indeed exist, then the following

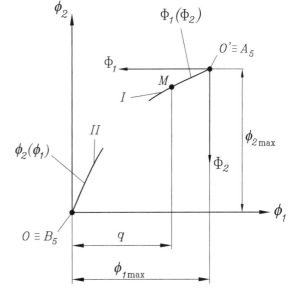

Figure 4.5.2. Illustration of direct function $\phi_2(\phi_1)$ and inverted function $\Phi_1(\Phi_2)$.

conditions must be observed:

$$\phi_2(\phi_1) = \Phi_1(\Phi_2), \quad \phi_{2max} = \phi_{1max} \tag{4.5.1}$$

(2) Equation (4.5.1) confirms the following:

(a) Assume that the position of rotation of centrodes 1 and 2 corresponds to A_5, and rotation is provided from centrode 2 to centrode 1 with the transmission function $\Phi_1(\Phi_2)$.

(b) Due to the identity of centrodes, rotation of centrode 1 with respect to centrode 2 (at the position of rotation A_5) must be provided with observation of the equation $\Phi_1(\Phi_2) = \phi_2(\phi_1)$. Function $\phi_2(\phi_1)$ relates rotation of centrode 2 with respect to centrode 1 at the position of rotation B_5.

(c) Due to the identity of centrode 1 and 2, it must be observed as well that the maximal angles of rotation of centrodes 1 and 2 must be equal.

(3) Consider now the instantaneous positions of the centrodes that correspond to point M of the graph $M - A_5$ of function $\phi_2(\phi_1)$ (Fig. 4.5.2, branch I). It is easy to verify that $q = \phi_{1max} - \Phi_1(\Phi_2)$. However, because $\Phi_1(\Phi_2)$ and $\phi_2(\phi_1)$ are related by Eq. (4.5.1), we have

$$q = \phi_{1max} - \Phi_1(\Phi_2) = \phi_{1max} - \phi_2(\phi_1) = \phi_{1max} - F(\phi_1) \tag{4.5.2}$$

Function $F(q)$ is the same as $F(\phi_1) \equiv \phi_2(\phi_1)$, so then we obtain

$$F(q) = \phi_{2max} - \Phi_2(\Phi_1) \tag{4.5.3}$$

Functions $\Phi_2(\Phi_1)$ and $\phi_1(\phi_2)$ are the direct and inverted ones. We recall as well that $\phi_{2max} = \phi_{1max}$. We then obtain

$$F(q) = \phi_{2max} - \Phi_2(\Phi_1) = \phi_{2max} - \phi_1(\phi_2) = \phi_{1max} - \phi_1(\phi_2) \tag{4.5.4}$$

Using Eqs. (4.5.2) and (4.5.3), we obtain

$$q = \phi_{1max} - F(\phi_1) \tag{4.5.5}$$

$$F(q) = \phi_{1max} - \phi_1 \tag{4.5.6}$$

It follows from Eq. (4.5.5), and the meaning from operation $F(q)$, that

$$F(q) = F\left(\phi_{1max} - F(\phi_1)\right) \tag{4.5.7}$$

Equations (4.5.5) and (4.5.7) yield the following final expression of the sought-for functional for identical centrodes:

$$F\left(\phi_{1max} - F(\phi_1)\right) = \phi_{1max} - \phi_1 \tag{4.5.8}$$

We describe the application of Eq. (4.5.8) for the case of identical centrodes of a drive that generates function $y = \dfrac{1}{x}$.

PROBLEM 4.5.1. This example represents the design of noncircular gears for generation of function $y = \dfrac{1}{x}$ in the interval $x_1 \leq x \leq x_2$. The design of the gear centrodes

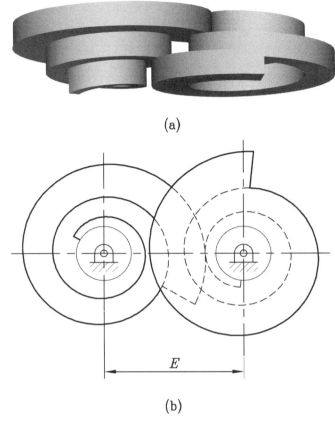

(a)

(b)

Figure 4.5.3. Identical unclosed centrodes, for generation of $y = \dfrac{1}{x}$, $1 < x < 3$, $\phi_{1max} = \phi_{2max} = 5\pi$ (a) in 3D-space; (b) in 2D-space.

is based on

$$\phi_1 = k_1(x - x_1) \tag{4.5.9}$$

$$\phi_2 = k_2(y_1 - y) = k_2\left(\frac{1}{x_1} - \frac{1}{x}\right) \tag{4.5.10}$$

where

$$k_1 = \frac{\phi_{1max}}{x_2 - x_1}, \quad k_2 = \frac{\phi_{2max}}{\frac{1}{x_1} - \frac{1}{x_2}}, \quad (\phi_{2max} = \phi_{1max}) \tag{4.5.11}$$

Transmission function $F(\phi_1)$ is

$$F(\phi_1) \equiv \phi_2(\phi_1) = \frac{k_2\phi_1}{x_1\phi_1 + k_1x_1^2} = \frac{a_2\phi_1}{a_3 + a_4\phi_1} \tag{4.5.12}$$

where

$$a_2 = k_2, \quad a_3 = k_1x_1^2, \quad a_4 = x_1 \tag{4.5.13}$$

The condition $\phi_{2max} = \phi_{1max}$ of the identity of centrodes is observed; thus, $\phi_{1max} = \dfrac{a_2 - a_3}{a_4}$.

The Functional (4.5.8) for the considered example of design is represented as

$$\frac{a_2\big(\phi_{1max} - F(\phi_1)\big)}{a_3 + a_4\big(\phi_{1max} - F(\phi_1)\big)} = \phi_{1max} - \phi_1 \qquad (4.5.14)$$

Observation of this equation is the proof of the identity of centrode 1 and 2 of the designed gear drive.

The centrodes of the gears are represented by

$$r_1(\phi_1) = \frac{E}{m_{12}(\phi_1) + 1} = \frac{a_2 a_3 E}{(a_3 + a_4\phi_1)^2 + a_2 a_3} \qquad (4.5.15)$$

where $m_{12}(\phi_1)$ is the derivative function,

$$r_2(\phi_1) = E - r_1(\phi_1) = \frac{E(a_3 + a_4\phi_1)^2}{(a_3 + a_4\phi_1)^2 + a_2 a_3} \qquad (4.5.16)$$

$$\phi_2(\phi_1) = F(\phi_1) = \frac{a_2\phi_1}{a_3 + a_4\phi_1} \qquad (4.5.17)$$

The centrodes are represented in Fig. 4.5.3.

5 Generation of Planar and Helical Elliptical Gears

5.1 Introduction

This chapter covers generation of planar and helical noncircular gears by a rack cutter, a hob, and a shaper. The cornerstone for determination of the generated tooth surfaces is the derivation of the equation of meshing of the generating and generated tooth surfaces. In this chapter, a matrix approach that allows the determination of relative velocity and equation of meshing to be simplified and computerized is presented. The developed theory will be illustrated with numerical examples of generation of planar and elliptical gears.

5.2 Generation of Elliptical Gears by Rack Cutter

ALGORITHM OF ROLLING MOTIONS. The main idea of the enveloping method for generation of planar noncircular gears for the case of application of a rack cutter is described as follows (Fig. 5.2.1):

(i) Centrode 3 of the rack cutter is a straight line $t - t$ that is a common tangent to centrodes 1 and 2 of mating noncircular gears 1 and 2 and rolls over 1 and 2.
(ii) Rolling is provided wherein the rack cutter translates along tangent $t - t$ and is rotated about the instantaneous center of rotation I.
(iii) Tooth surfaces of gear 1, gear 2, and rack cutter 3 are in mesh simultaneously, and gear 1 and 2 are provided with conjugated surfaces.

The related motions of the rack cutter and the noncircular gear may be determined considering the motions of the generating tool and one of the centrodes of the gear set as follows (Fig. 5.2.2):

(a) The rack cutter centrode (denoted as I) is in mesh with centrode II of a noncircular gear.
(b) Rolling is provided by observation of

$$\mathbf{v}^{(I)} = \mathbf{v}_{\text{rot}}^{(II)} + \mathbf{v}_{\text{tr}}^{(II)} \tag{5.2.1}$$

71

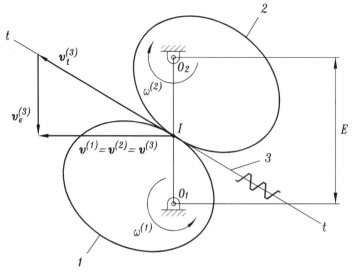

Figure 5.2.1. Illustration of generation of noncircular gears 1 and 2 by rack cutter 3.

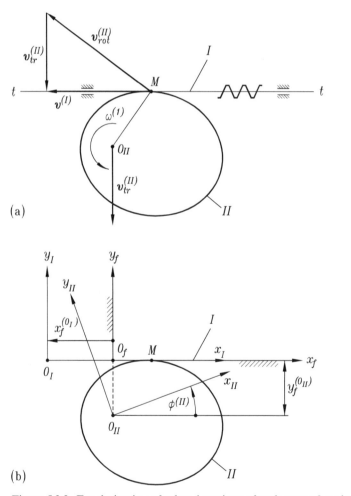

(a)

(b)

Figure 5.2.2. For derivation of related motions of rack cutter 2 and noncircular gear 1.

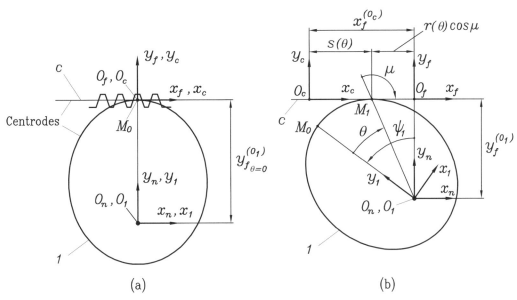

Figure 5.2.3. For derivation of algorithm of rolling motions: (a) initial position of centrodes c of rack cutter and 1 of noncircular gear; (b) current position of centrodes c and 1.

Equation (5.2.1) is obtained by considering that the rack cutter performs only translational motion with velocity $\mathbf{v}^{(I)}$ along the common tangent to centrodes I and II. The noncircular gear II performs rotational motion about center of rotation O_{II}, and translational motion in the direction that is perpendicular to $t - t$. Vectors $\mathbf{v}_{\text{rot}}^{(II)}$ and $\mathbf{v}_{\text{tr}}^{(II)}$ represent velocities of gear II of rotational and translational motions (Fig. 5.2.2(a)). Figure 5.2.2(b) illustrates the positions of the rack cutter I and noncircular gear II in fixed coordinate system S_f. It will be shown that functions $x_f^{(O_I)}(\phi^{(II)})$ and $y_f^{(O_{II})}(\phi^{(II)})$ are nonlinear.

The derivation of the algorithm of rolling motions is based on application of the following planar coordinate systems (Figs. 5.2.3(a) and (b)): (i) movable coordinate systems S_c and S_1 rigidly connected to the rack cutter and the noncircular gear, respectively, (ii) a fixed coordinate system S_f, and (iii) an auxiliary movable coordinate system S_n that performs translation along axis y_f (Fig. 5.2.3(b)).

Coordinate systems S_c, S_1, S_f, and S_n are planar; however, the developed algorithm of rolling developed for generation of planar noncircular gears may be applied for generation of helical noncircular gears as well (see following). The derivations are discussed for the case of elliptical gears. However, the obtained results may be applied for other types of noncircular gears.

Figure 5.2.3(a) and (b) show respectively the initial and current positions of the centrodes of the rack cutter c and the noncircular gear 1. The following is the preliminary information about the centrodes mentioned here and their motions performed during the meshing.

(1) The rack cutter centrode c is a straight line directed along axes x_c, x_f (Fig. 5.2.3(b)) that performs translational motion along x_f.

(2) The centrode of noncircular gear 1 is an ellipse represented in polar form, according to Eq. (4.2.9), by

$$r(\theta) = \frac{a(1 - e^2)}{1 - e \cos \theta} \tag{5.2.2}$$

Here, a is the major axis of the ellipse and e is the eccentricity of the ellipse (see Section 4.2.2). The polar axis of the ellipse is y_1 (Fig. 5.2.3(b)). The ellipse centrode (during the meshing with the centrode of the rack cutter) performs:
  (i) Rotation about O_1 (it is the focus of the ellipse), and
  (ii) Translational motion along axis y_n (Fig. 5.2.3(b)).
(3) The instantaneous point of tangency of centrode c and 1 is M_1 (Fig. 5.2.3(b)). Because centrode c and 1 roll over each other, the length of the arc $M_0 - M_1$ is equal to the distance $|\overline{O_c M_1}|$.
(4) Parameter ψ_1 determines the angular position of coordinate system $S_1(x_1, y_1)$ with respect to $S_f(x_f, y_f)$.
(5) Parameter μ (Fig. 5.2.3(b)) determines orientation of the tangent to ellipse 1 at point M_1, where (see Section 2.5)

$$\tan \mu = \frac{r(\theta)}{\dfrac{dr}{d\theta}} \tag{5.2.3}$$

Observation of conditions of pure rolling yields the following relations between the motions of the rack cutter c and noncircular gear 1:

$$x_f^{(O_c)}(\theta) = -s(\theta) + r(\theta) \cos \mu = -\int_0^\theta \frac{r(\theta)}{\sin \mu} d\theta + r(\theta) \cos \mu \tag{5.2.4}$$

$$y_f^{(O_1)}(\theta) = -r(\theta) \sin \mu \tag{5.2.5}$$

$$\psi_1(\theta) = \theta + \mu - \mu_0, \qquad \mu_0 = \frac{\pi}{2} \tag{5.2.6}$$

$$\mu(\theta) = \arctan \left(-\frac{1 - e \cos \theta}{e \sin \theta} \right) \tag{5.2.7}$$

For the purpose of derivation of the equation of meshing, the derivation of the functions mentioned here is required. The results of derivation of Eqs. (5.2.4)–(5.2.7) with respect to polar angle θ are as follows:
The derivative of polar vector $r(\theta)$ is given by

$$\frac{dr(\theta)}{d\theta} = \frac{d}{d\theta} \frac{p}{1 - e \cos \theta} = \frac{-pe \sin \theta}{(1 - e \cos \theta)^2} \tag{5.2.8}$$

The derivative of angle $\mu(\theta)$ is given by

$$\frac{d\mu(\theta)}{d\theta} = \frac{d}{d\theta} \arctan \frac{-1 + e \cos \theta}{e \sin \theta} = \frac{-e(e - \cos \theta)}{1 + e(e - 2 \cos \theta)} \tag{5.2.9}$$

The derivative of function $s(\theta)$ is given by

$$\frac{ds(\theta)}{d\theta} = \frac{d}{d\theta}\int_0^\theta \frac{r(\theta)}{\sin\mu(\theta)}d\theta = \frac{r(\theta)}{\sin\mu(\theta)} \qquad (5.2.10)$$

The derivative of magnitude $x_f^{(O_c)}(\theta)$ is then given by

$$\begin{aligned}
\frac{dx_f^{(O_c)}}{d\theta} &= \frac{d}{d\theta}[-s(\theta) + r(\theta)\cos\mu(\theta)] \\
&= -\frac{ds(\theta)}{d\theta} + \frac{dr(\theta)}{d\theta}\cos\mu(\theta) - r(\theta)\sin\mu(\theta)\frac{d\mu(\theta)}{d\theta} \qquad (5.2.11)
\end{aligned}$$

The derivative of magnitude $y_f^{(O_1)}(\theta)$ is given by

$$\frac{dy_f^{(O_1)}}{d\theta} = \frac{d}{d\theta}[-r(\theta)\sin\mu(\theta)] = -\frac{dr(\theta)}{d\theta}\sin\mu(\theta) - r(\theta)\cos\mu(\theta)\frac{d\mu(\theta)}{d\theta}$$

$$(5.2.12)$$

The derivative of magnitude $\psi_1(\theta)$ is given by

$$\frac{d\psi_1}{d\theta} = \frac{d}{d\theta}[\theta + \mu - \frac{\pi}{2}] = 1 + \frac{d\mu(\theta)}{d\theta} \qquad (5.2.13)$$

DERIVATION OF RACK CUTTER GENERATING SURFACES REPRESENTED IN COORDINATE SYSTEM S_c. Two types of generating surfaces are considered: (i) a planar rack cutter applied for generation of planar gears, represented in coordinate system S_t (Fig. 5.2.4(a)), and (ii) a skew rack cutter that is formed by the motion of S_t in coordinate system S_c, as illustrated in Fig. 5.2.4(b) and (c).

Surface Σ_c of the skew rack cutter and the surface unit normal are represented in S_c considering the coordinate transformation from S_t to S_c. The position vector $\mathbf{r}_c(u_c, l_c)$ and surface normal $\mathbf{n}_c$ of Σ_c are determined as

$$\mathbf{r}_c(u_c, l_c) = \begin{bmatrix} \left(\pm u_c\sin\alpha_c \mp \dfrac{\pi m_n}{4}\right)\cos\beta_c + l_c\sin\beta_c \\ u\cos\alpha_c \\ \left(\mp u_c\sin\alpha_c \pm \dfrac{\pi m_n}{4}\right)\sin\beta_c + l_c\cos\beta_c \\ 1 \end{bmatrix} \qquad (5.2.14)$$

$$\mathbf{n}_c = \begin{bmatrix} \mp\cos\alpha_c\cos\beta_c \\ \sin\alpha_c \\ \pm\cos\alpha_c\sin\beta_c \end{bmatrix} \qquad (5.2.15)$$

Equation (5.2.14) represents in S_c the skew rack cutter Σ_c as a *plane* with surface parameters (u_c, l_c). The unit normal $\mathbf{n}_c$ is represented by Eq. (5.2.15). The upper and lower signs in Eqs. (5.2.14) and (5.2.15) correspond to the left and right sides of Σ_c.

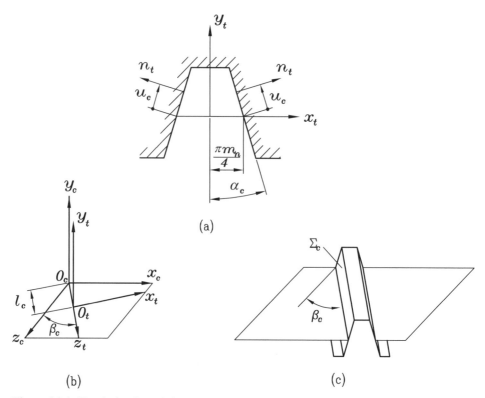

Figure 5.2.4. For derivation of planar and skew rack cutter: (a) representation of planar rack cutter in coordinate system S_t; (b) toward representation of skew rack cutter in coordinate system S_c; (c) representation of skew rack cutter.

Taking in Eqs. (5.2.14) and (5.2.15) $\beta_c = 0$, we obtain the planar rack-cutter surface and its unit normal for generation of planar elliptical gears.

GENERATION OF SURFACE Σ_1 OF ELLIPTICAL GEAR BY RACK CUTTER. The derivation of Σ_1 is based on the following steps:

(i) Determination of coordinate transformation between coordinate systems S_c and S_1 that allows the family of generating rack-cutter surfaces to be represented in coordinate system S_1 in three-parametric form as

$$\mathbf{r}_1(u_c, l_c, \theta) = \mathbf{M}_{1c}(\theta)\mathbf{r}_c(u_c, l_c) \qquad (5.2.16)$$

where matrix $\mathbf{M}_{1c}$ represents the coordinate transformation in procedures applied from S_c to S_1.

(ii) Derivation of equation of meshing

$$f_{1c}(u_c, l_c, \theta) = 0 \qquad (5.2.17)$$

that relates parameters (u_c, l_c, θ). Simultaneous consideration of Eqs. (5.2.16) and (5.2.17) determines surface Σ_1 of helical elliptical gear generated by the skew rack cutter.

COORDINATE TRANSFORMATION IN TRANSITION FROM COORDINATE SYSTEM S_c TO S_1. Derivation of Eq. (5.2.16) is performed as follows:

$$\mathbf{r}_1(u_c, l_c, \theta) = \mathbf{M}_{1c}(\theta)\mathbf{r}_c(u_c, l_c) = \mathbf{M}_{1n}(\theta)\mathbf{M}_{nf}(\theta)\mathbf{M}_{fc}(\theta)\mathbf{r}_c(u_c, l_c)$$

$$= \begin{bmatrix} \cos\psi_1 & \sin\psi_1 & 0 & 0 \\ -\sin\psi_1 & \cos\psi_1 & 0 & 0 \\ 0 & 0 & 1 & 0 \\ 0 & 0 & 0 & 1 \end{bmatrix} \begin{bmatrix} 1 & 0 & 0 & 0 \\ 0 & 1 & 0 & -y_f^{(O_1)} \\ 0 & 0 & 1 & 0 \\ 0 & 0 & 0 & 1 \end{bmatrix} \begin{bmatrix} 1 & 0 & 0 & x_f^{(O_c)} \\ 0 & 1 & 0 & 0 \\ 0 & 0 & 1 & 0 \\ 0 & 0 & 0 & 1 \end{bmatrix} \mathbf{r}_c(u_c, l_c)$$

$$= \begin{bmatrix} \cos\psi_1 & \sin\psi_1 & 0 & x_f^{(O_c)}\cos\psi_1 - y_f^{(O_1)}\sin\psi_1 \\ -\sin\psi_1 & \cos\psi_1 & 0 & -x_f^{(O_c)}\sin\psi_1 - y_f^{(O_1)}\cos\psi_1 \\ 0 & 0 & 1 & 0 \\ 0 & 0 & 0 & 1 \end{bmatrix} \mathbf{r}_c(u_c, l_c) \qquad (5.2.18)$$

MATRIX DERIVATION OF EQUATION OF MESHING $f_{1c}(u_c, l_c, \theta) = 0$. We consider that the rack cutter is provided by generating surface Σ_c represented in 3D space by vector function $\mathbf{r}_c(u_c, l_c)$. Using coordinate transformation from coordinate system S_c to S_1, we obtain in S_1 a family of generating surfaces $\mathbf{r}_1(u_c, l_c, \theta)$.

The generated surface of the noncircular gear is determined as the envelope to the family of surfaces $\mathbf{r}_1(u_c, l_c, \theta)$. In differential geometry (for instance, see Korn & Korn, 1968) the envelope to the family of surfaces $\mathbf{r}_1(u_c, l_c, \theta)$ can be determined by simultaneous consideration of vector function $\mathbf{r}_1(u_c, l_c, \theta)$ and the equation of meshing:

$$f(u_c, l_c, \theta) = \left(\frac{\partial \mathbf{r}_1}{\partial u_c} \times \frac{\partial \mathbf{r}_1}{\partial l_c} \right) \cdot \frac{\partial \mathbf{r}_1}{\partial \theta} = 0 \qquad (5.2.19)$$

A matrix approach for derivation of equation of meshing that allows us to simplify and computerize the determination of the surface generated by the rack cutter has been developed, and is performed as follows.

The family of generating rack-cutter surfaces is represented in coordinate system S_1 in three-parametric form by Eq. (5.2.18). Considering Cartesian coordinates instead of homogeneous coordinates, the gear tooth surface may be obtained as

$$\boldsymbol{\rho}_1(u_c, l_c, \theta) = \begin{bmatrix} \cos\psi_1 & \sin\psi_1 & 0 \\ -\sin\psi_1 & \cos\psi_1 & 0 \\ 0 & 0 & 1 \end{bmatrix} \boldsymbol{\rho}_c(u_c, l_c) + \begin{bmatrix} \cos\psi_1 & \sin\psi_1 & 0 \\ -\sin\psi_1 & \cos\psi_1 & 0 \\ 0 & 0 & 1 \end{bmatrix} \begin{bmatrix} x_f^{(O_c)} \\ -y_f^{(O_1)} \\ 0 \end{bmatrix}$$

$$= \mathbf{L}_{1c}\boldsymbol{\rho}_c + \mathbf{L}_{1c}\mathbf{R} \qquad (5.2.20)$$

Here,

$$\mathbf{R} = \overline{\mathbf{O_n O_c}} = \begin{bmatrix} x_f^{(O_c)} & -y_f^{(O_1)} & 0 \end{bmatrix}^T \tag{5.2.21}$$

The proposed approach allows us to obtain (i) the relative velocity of rack cutter c with respect to gear 1

$$\mathbf{v}_1^{(c1)} = \frac{d\boldsymbol{\rho}_1}{dt} = \dot{\mathbf{L}}_{1c}\boldsymbol{\rho}_c + \dot{\mathbf{L}}_{1c}\mathbf{R} + \mathbf{L}_{1c}\dot{\mathbf{R}} \tag{5.2.22}$$

(ii) and the relative velocity of gear 1 with respect to rack cutter c

$$\mathbf{v}_c^{(1c)} = -\mathbf{L}_{c1}\mathbf{v}_1^{(c1)} = -\mathbf{L}_{c1}\left(\dot{\mathbf{L}}_{1c}\boldsymbol{\rho}_c + \dot{\mathbf{L}}_{1c}\mathbf{R} + \mathbf{L}_{1c}\dot{\mathbf{R}} \right) \tag{5.2.23}$$

The final expression of equation of meshing is

$$f_1^{(c1)}(u_c, l_c, \theta) = \left(\mathbf{L}_{1c}\mathbf{n}_c^{(c)} \right) \cdot \left(\dot{\mathbf{L}}_{1c}\boldsymbol{\rho}_c + \dot{\mathbf{L}}_{1c}\mathbf{R} + \mathbf{L}_{1c}\dot{\mathbf{R}} \right) = 0 \tag{5.2.24}$$

or

$$f_c^{(1c)}(u_c, l_c, \theta) = \mathbf{n}_c^{(c)} \cdot \left[-\mathbf{L}_{c1}\left(\dot{\mathbf{L}}_{1c}\boldsymbol{\rho}_c + \dot{\mathbf{L}}_{1c}\mathbf{R} + \mathbf{L}_{1c}\dot{\mathbf{R}} \right) \right] = 0 \tag{5.2.25}$$

where

$$\dot{\mathbf{L}}_{1c} = \begin{bmatrix} -\sin\psi_1 & \cos\psi_1 & 0 \\ -\cos\psi_1 & -\sin\psi_1 & 0 \\ 0 & 0 & 0 \end{bmatrix} \dot{\psi}_1 \tag{5.2.26}$$

$$\dot{\mathbf{R}} = \overline{\dot{\mathbf{O_n O_c}}} = \begin{bmatrix} \dot{x}_f^{(O_c)} & -\dot{y}_f^{(O_1)} & 0 \end{bmatrix}^T \tag{5.2.27}$$

Derivatives $\dot{x}_f^{(O_c)}$, $\dot{y}_f^{(O_1)}$, and $\dot{\psi}_1$ are

$$\dot{x}_f^{(O_c)} = \frac{dx_f^{(O_c)}}{d\theta} \frac{d\theta}{dt} \tag{5.2.28}$$

$$\dot{y}_f^{(O_1)} = \frac{dy_f^{(O_1)}}{d\theta} \frac{d\theta}{dt} \tag{5.2.29}$$

$$\dot{\psi}_1 = \frac{d\psi_1}{d\theta} \frac{d\theta}{dt} \tag{5.2.30}$$

where $\dfrac{dx_f^{(O_c)}}{d\theta}$, $\dfrac{dy_f^{(O_1)}}{d\theta}$, and $\dfrac{d\psi_1}{d\theta}$ are given by Eqs. (5.2.11), (5.2.12), and (5.2.13), respectively.

The tooth surface of the helical elliptical gear is determined by simultaneous consideration of Eq. (5.2.20) with Eq. (5.2.24) or (5.2.25).

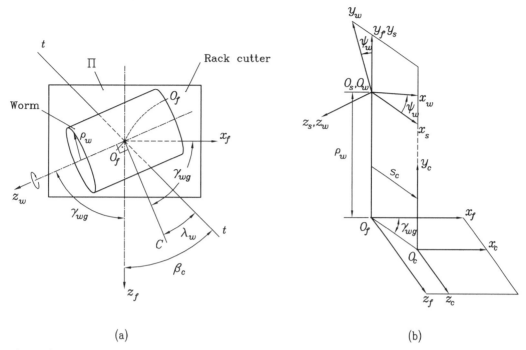

Figure 5.3.1. For determination of the worm: (a) installation of worm with respect to rack cutter; (b) applied coordinate systems.

5.3 Generation of Elliptical Gears by Hob

DERIVATION OF WORM-THREAD GENERATING SURFACES REPRESENTED IN COORDINATE SYSTEM S_w. Application of a grinding worm or a hob for generation of elliptical gears is useful for an improvement of productivity and reliability of generated gears.

A worm-thread surface Σ_w, that is in imaginary meshing with rack-cutter tooth surface Σ_c, is being determined. Conditions of meshing between both surfaces, Σ_w and Σ_c, allows the worm-thread surface to be determined. The procedure is as follows:

(1) Worm and gear shafts form crossing angle γ_{wg}. Figure 5.3.1(a) shows the installation of worm axode on rack-cutter axode plane Π. Angle γ_{wg} is given as

$$\gamma_{wg} = \frac{\pi}{2} - \beta_c + \lambda_w \tag{5.3.1}$$

wherein β_c is the helix angle of the skew rack cutter and λ_w is the lead angle of the worm.
(2) Line $t - t$ at the pitch point O_f, and the helix of the gear (see Fig. 5.3.1(a)), are in tangency.
(3) Moveable coordinate systems S_c and S_w are rigidly connected to surfaces Σ_c and Σ_w, respectively. Fixed coordinate systems S_f and S_s are considered for

definition of motions of rack-cutter and worm tooth surfaces, respectively (see Fig. 5.3.1(b)).

(4) Surface Σ_c is considered as given and performs the following motions:

(i) Translation s_c of coordinate system S_c along the straight line $O_f - C$ (see Fig. 5.3.1(b)). Straight line $O_f - C$ forms angle γ_{wg} with respect to axis x_f (see Fig. 5.3.1(a)).

(ii) Rotation of coordinate system S_w about axis z_s an angle ψ_w, determined as

$$\psi_w = \frac{s_c}{\rho_w} \tag{5.3.2}$$

wherein ρ_w is the pitch radius of the griding worm.

(5) Worm-thread surface Σ_w is obtained by simultaneous consideration of vector equation

$$\mathbf{r}_w(u_c, l_c, \psi_w) = \mathbf{M}_{wc}(\psi_w)\mathbf{r}_c(u_c, l_c) \tag{5.3.3}$$

and equation of meshing

$$\left(\frac{\partial \mathbf{r}_c}{\partial u_c} \times \frac{\partial \mathbf{r}_c}{\partial l_c}\right) \cdot \frac{\partial \mathbf{r}_c}{\partial \psi_w} = 0 \tag{5.3.4}$$

Here, ψ_w is the generalized parameter of meshing, and $\mathbf{M}_{wc}$ is a 4×4 matrix that represents coordinate transformation of homogeneous coordinates from system S_c to system S_w. However, we will apply for derivation of Eq. (5.3.4), the matrix approach presented for the case of rack-cutter generation.

Worm-thread surface Σ_w may be considered for the purpose of simplicity as a surface with two independent parameters (h_w, ε_w).

GENERATION OF SURFACE Σ_1 OF ELLIPTICAL GEAR BY A WORM. The derivation of Σ_1 by a worm-thread surface Σ_w is based on the following procedure (see Fig. 5.3.2):

(i) Two coordinate systems S_w and S_1 are rigidly connected to the worm-thread surface and the to-be-determined gear-tooth surface. A fixed reference system S_f is considered for definition of motions of systems S_w and S_1.

(ii) Worm-thread surface Σ_w is given by vector function $\mathbf{R}_w(h_w, \varepsilon_w)$.

(iii) Two sets of motions are provided to the worm:

(a) Rotation of the worm about axis z_w on angle ϕ_w.

(b) Translation s_w along axis z_f that is parallel to the axis of the gear. Coordinate system S_s is a movable coordinate system that is translated with system S_w.

(iv) Rotation and translation of the worm are accompanied by rotation and translation of the elliptical gear as follows:

(a) Rotation ψ_1 about axis z_1 of the gear.

(b) Translation defined by position $y_f^{(O_1)}$ along axis y_f.

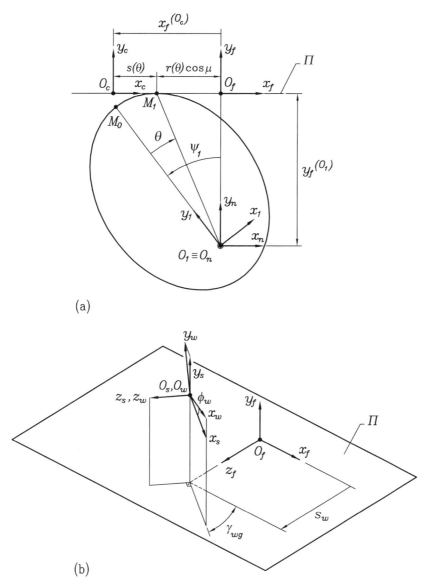

Figure 5.3.2. Applied coordinate systems for generation of the elliptical gear by a grinding worm: (a) at the elliptical gear; (b) at the worm.

Magnitudes ψ_1 and $y_f^{(O_1)}$ may be determined as functions of polar angle θ (see Eqs. (5.2.5) and (5.2.6)). Magnitudes ψ_1 and $y_f^{(O_1)}$ are related with motions ϕ_w and s_w by functions

$$y_f^{(O_1)}(\theta) = -r(\theta)\sin\mu \tag{5.3.5}$$

$$\psi_1(\theta) = \theta + \mu - \mu_0, \qquad \mu_0 = \frac{\pi}{2} \tag{5.3.6}$$

and the new function to be determined as

$$g(\phi_w, s_w, \theta) = 0 \tag{5.3.7}$$

(v) Coordinate transformation between systems S_w and S_1 determines the family of worm-thread surfaces in system S_1 as

$$\mathbf{r}_1(h_w, \varepsilon_w, \phi_w, s_w) = \mathbf{M}_{1w}(\phi_w, s_w)\mathbf{R}_w(h_w, \varepsilon_w) \qquad (5.3.8)$$

Here, ϕ_w and s_w are independent generalized parameters of motion, which means that the generation is a double-enveloping process, and 4×4 matrix $\mathbf{M}_{1w}$ describes coordinate transformation from system S_w to system S_1. Derivation of Eq. (5.3.8) is performed as

$$\mathbf{r}_1(h_w, \varepsilon_w, \phi_w, s_w) = \mathbf{M}_{1f}(\phi_w, s_w)\mathbf{M}_{fs}(s_w)\mathbf{M}_{sw}(\phi_w)\mathbf{R}_w(h_w, \varepsilon_w) \qquad (5.3.9)$$

wherein

$$\mathbf{M}_{1f}(\phi_w, s_w) = \begin{bmatrix} \cos\psi_1 & \sin\psi_1 & 0 & -y_f^{(O_1)}\sin\psi_1 \\ -\sin\psi_1 & \cos\psi_1 & 0 & -y_f^{(O_1)}\cos\psi_1 \\ 0 & 0 & 1 & 0 \\ 0 & 0 & 0 & 1 \end{bmatrix} \qquad (5.3.10)$$

$$\mathbf{M}_{fs}(s_w) = \begin{bmatrix} \cos\gamma_{wg} & 0 & -\sin\gamma_{wg} & 0 \\ 0 & 1 & 0 & \rho_w \\ \sin\gamma_{wg} & 0 & \cos\gamma_{wg} & s_w \\ 0 & 0 & 0 & 1 \end{bmatrix} \qquad (5.3.11)$$

$$\mathbf{M}_{sw}(\phi_w) = \begin{bmatrix} \cos\phi_w & -\sin\phi_w & 0 & 0 \\ \sin\phi_w & \cos\phi_w & 0 & 0 \\ 0 & 0 & 1 & 0 \\ 0 & 0 & 0 & 1 \end{bmatrix} \qquad (5.3.12)$$

(vi) Because generation of surface Σ_1 of the elliptical gear involves a double-enveloping process, there are two equations of meshing,

$$f_1^{(w1)}(h_w, \varepsilon_w, \phi_w, s_w) = 0 \qquad (5.3.13)$$

$$f_2^{(w1)}(h_w, \varepsilon_w, \phi_w, s_w) = 0 \qquad (5.3.14)$$

that relate parameters $(h_w, \varepsilon_w, \phi_w, s_w)$. Simultaneous consideration of Eqs. (5.3.8), (5.3.13), and (5.3.14) determine surface Σ_1 of the elliptical gear.

DERIVATION OF FUNCTION $g(\phi_w, s_w, \theta) = 0$ The derivation is performed as follows (see Fig. 5.3.3):

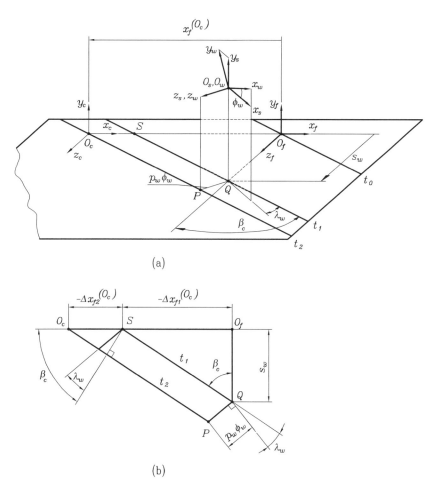

Figure 5.3.3. For determination of function $g(\phi_w, s_w, \theta) = 0$.

(i) An imaginary rack cutter is considered being in simultaneous meshing with the elliptical gear and the worm. At the initial position, system S_c coincides with system S_f, and common tangent line $t - t$ between the three surfaces, Σ_w, Σ_1, and Σ_c, is at position t_0.

(ii) Due to rotation and translation of the worm on ϕ_w and s_w, the common tangent $t - t$ will take position t_2. The location of system S_c in S_f is determined by $x_f^{(O_c)}$.

(iii) Displacement of system S_c may be obtained as the sum of independent displacements $\Delta x_{f1}^{(O_c)}$ and $\Delta x_{f2}^{(O_c)}$:

$$\Delta x_f^{(O_c)} = \Delta x_{f1}^{(O_c)} + \Delta x_{f2}^{(O_c)} \tag{5.3.15}$$

Displacement $\Delta x_{f1}^{(O_c)} = \overline{O_f S}$ is caused by translation s_w and is defined by positions t_0 and t_1. Displacement $\Delta x_{f2}^{(O_c)} = \overline{S O_c}$ is caused by rotation ϕ_w and is defined by positions t_1 and t_2.

(iv) Illustrations of Fig. 5.3.3(b) yield

$$-\Delta x_{f1}^{(O_c)} = \overline{O_f S} = \tan\beta_c s_w \tag{5.3.16}$$

$$-\Delta x_{f2}^{(O_c)} = \overline{SO_C} = \frac{p_w\cos\lambda_w}{\cos\beta_c}\phi_w \tag{5.3.17}$$

where p_w is the pitch of the worm.

(v) Because $x_f^{(O_c)}$ depends on polar angle θ (see Eq. (5.2.4)), function $g(\phi_w, s_w, \theta) = 0$ is obtained finally as

$$g(\phi_w, s_w, \theta) = x_f^{(O_c)}(\theta) + \tan\beta_c s_w + \frac{p_w\cos\lambda_w}{\cos\beta_c}\phi_w = 0 \tag{5.3.18}$$

MATRIX DERIVATION OF EQUATIONS OF MESHING. Matrix derivation of equations of meshing $f_1^{(w1)}(h_w, \varepsilon_w, \phi_w, s_w) = 0$ and $f_2^{(w1)}(h_w, \varepsilon_w, \phi_w, s_w) = 0$ is performed as follows:

(i) Vector position $\mathbf{r}_1(h_w, \varepsilon_w, \phi_w, s_w)$ in homogeneous coordinates is given by

$$\mathbf{r}_1(h_w, \varepsilon_w, \phi_w, s_w) = \mathbf{M}_{1w}(\phi_w, s_w)\mathbf{R}_w(h_w, \varepsilon_w)$$

$$= \begin{bmatrix} a_{11} & a_{12} & a_{13} & \sin\psi_1(\rho_w - y_f^{(O_1)}) \\ a_{21} & a_{22} & a_{23} & \cos\psi_1(\rho_w - y_f^{(O_1)}) \\ a_{31} & a_{32} & a_{33} & s_w \\ 0 & 0 & 0 & 1 \end{bmatrix}\mathbf{R}_w(h_w, \varepsilon_w) \tag{5.3.19}$$

wherein

$$a_{11} = \cos\psi_1\cos\gamma_{wg}\cos\phi_w + \sin\psi_1\sin\phi_w \tag{5.3.20}$$

$$a_{12} = -\cos\psi_1\cos\gamma_{wg}\sin\phi_w + \sin\psi_1\cos\phi_w \tag{5.3.21}$$

$$a_{13} = -\cos\psi_1\sin\gamma_{wg} \tag{5.3.22}$$

$$a_{21} = -\sin\psi_1\cos\gamma_{wg}\cos\phi_w + \cos\psi_1\sin\phi_w \tag{5.3.23}$$

$$a_{22} = \sin\psi_1\cos\gamma_{wg}\sin\phi_w + \cos\psi_1\cos\phi_w \tag{5.3.24}$$

$$a_{23} = \sin\psi_1\sin\gamma_{wg} \tag{5.3.25}$$

$$a_{31} = \sin\gamma_{wg}\cos\phi_w \tag{5.3.26}$$

$$a_{32} = -\sin\gamma_{wg}\sin\phi_w \tag{5.3.27}$$

$$a_{33} = \cos\gamma_{wg} \tag{5.3.28}$$

(ii) Vector position $\mathbf{r}_1(h_w, \varepsilon_w, \phi_w, s_w)$ is denoted as $\boldsymbol{\rho}_1(h_w, \varepsilon_w, \phi_w, s_w)$ in Cartesian coordinates and may be represented as

$$\boldsymbol{\rho}_1(h_w, \varepsilon_w, \phi_w, s_w) = \begin{bmatrix} a_{11} & a_{12} & a_{13} \\ a_{21} & a_{22} & a_{23} \\ a_{31} & a_{32} & a_{33} \end{bmatrix} \boldsymbol{\rho}_w(h_w, \varepsilon_w) + \begin{bmatrix} \sin\psi_1(\rho_w - y_f^{(O_1)}) \\ \cos\psi_1(\rho_w - y_f^{(O_1)}) \\ s_w \end{bmatrix}$$

$$= \mathbf{L}_{1w}(\phi_w, s_w)\boldsymbol{\rho}_w(h_w, \varepsilon_w) + \mathbf{R} \qquad (5.3.29)$$

Here, $\mathbf{L}_{1w}$ is a 3×3 matrix obtained by eliminating the last row and the last column of 4×4 matrix $\mathbf{M}_{1w}$. Matrix $\mathbf{L}_{1w}$ may be also obtained as

$$\mathbf{L}_{1w} = \mathbf{L}_{1f}\mathbf{L}_{fs}\mathbf{L}_{sw} \qquad (5.3.30)$$

whereas vector $\mathbf{R}$ is defined as

$$\mathbf{R} = \begin{bmatrix} \sin\psi_1(\rho_w - y_f^{(O_1)}) & \cos\psi_1(\rho_w - y_f^{(O_1)}) & s_w \end{bmatrix}^T \qquad (5.3.31)$$

(iii) Considering s_w as constant ($s_w = c$), the relative velocity of the worm-thread surface with respect to gear tooth surface may be obtained as

$$\mathbf{v}_{1,s_w=c}^{(w1)} = \dot{\boldsymbol{\rho}}_1 = \dot{\mathbf{L}}_{1w}\boldsymbol{\rho}_w + \dot{\mathbf{R}} \qquad (5.3.32)$$

wherein

$$\dot{\mathbf{L}}_{1w} = \dot{\mathbf{L}}_{1f}\mathbf{L}_{fs}\mathbf{L}_{sw} + \mathbf{L}_{1f}\mathbf{L}_{fs}\dot{\mathbf{L}}_{sw} \qquad (5.3.33)$$

$$\dot{\mathbf{L}}_{1f} = \begin{bmatrix} -\sin\psi_1 & \cos\psi_1 & 0 \\ -\cos\psi_1 & -\sin\psi_1 & 0 \\ 0 & 0 & 0 \end{bmatrix} \dot{\psi}_1 \qquad (5.3.34)$$

$$\dot{\mathbf{L}}_{sw} = \begin{bmatrix} -\sin\phi_w & -\cos\phi_w & 0 \\ \cos\phi_w & -\sin\phi_w & 0 \\ 0 & 0 & 0 \end{bmatrix} \dot{\phi}_w, \quad \dot{\phi}_w = -\frac{\cos\beta_c}{p_w\cos\lambda_w}\dot{x}_f^{(O_c)} \qquad (5.3.35)$$

$$\dot{\mathbf{R}} = \begin{bmatrix} \cos\psi_1(\rho_w - y_f^{(O_1)})\dot{\psi}_1 - \sin\psi_1\dot{y}_f^{(O_1)} \\ -\sin\psi_1(\rho_w - y_f^{(O_1)})\dot{\psi}_1 - \cos\psi_1\dot{y}_f^{(O_1)} \\ 0 \end{bmatrix} \qquad (5.3.36)$$

Then, the equation of meshing may be obtained as

$$f_1^{(w1)} = \mathbf{n}_1 \cdot \mathbf{v}_{1,s_w=c}^{(w1)} = 0 \qquad (5.3.37)$$

wherein

$$\mathbf{n}_1 = \mathbf{L}_{1w}\mathbf{n}_w \qquad (5.3.38)$$

Here, $\mathbf{n}_w$ is the unit normal to the worm-thread surface.

(iv) Considering ϕ_w as constant ($\phi_w = c$), the relative velocity of the worm-thread surface with respect to gear-tooth surface may be obtained as

$$\mathbf{v}_{1,\phi_w=c}^{(w1)} = \dot{\boldsymbol{\rho}}_1 = \dot{\mathbf{L}}_{1w}\rho_w + \dot{\mathbf{R}} \tag{5.3.39}$$

wherein

$$\dot{\mathbf{L}}_{1w} = \dot{\mathbf{L}}_{1f}\mathbf{L}_{fs}\mathbf{L}_{sw} \tag{5.3.40}$$

$$\dot{\mathbf{L}}_{1f} = \begin{bmatrix} -\sin\psi_1 & \cos\psi_1 & 0 \\ -\cos\psi_1 & -\sin\psi_1 & 0 \\ 0 & 0 & 0 \end{bmatrix}\dot{\psi}_1 \tag{5.3.41}$$

$$\dot{\mathbf{R}} = \begin{bmatrix} \cos\psi_1(\rho_w - y_f^{(O_1)})\dot{\psi}_1 - \sin\psi_1\dot{y}_f^{(O_1)} \\ -\sin\psi_1(\rho_w - y_f^{(O_1)})\dot{\psi}_1 - \cos\psi_1\dot{y}_f^{(O_1)} \\ \dot{s}_w \end{bmatrix} \tag{5.3.42}$$

$$\dot{s}_w = -\frac{1}{\tan\beta_c}\dot{x}_f^{(O_c)} \tag{5.3.43}$$

Then, the equation of meshing may be obtained as

$$f_2^{(w1)} = \mathbf{n}_1 \cdot \mathbf{v}_{1,\phi_w=c}^{(w1)} = 0 \tag{5.3.44}$$

(v) Derivations of derivatives $\dot{y}_f^{(O_1)}$, $\dot{x}_f^{(O_c)}$, and $\dot{\psi}_1$, are

$$\dot{y}_f^{(O_1)} = \frac{dy_f^{(O_1)}}{d\theta}\frac{d\theta}{dt} \tag{5.3.45}$$

$$\dot{x}_f^{(O_c)} = \frac{dx_f^{(O_c)}}{d\theta}\frac{d\theta}{dt} \tag{5.3.46}$$

$$\dot{\psi}_1 = \frac{d\psi_1}{d\theta}\frac{d\theta}{dt} \tag{5.3.47}$$

where $\dfrac{dx_f^{(O_c)}}{d\theta}$, $\dfrac{dy_f^{(O_1)}}{d\theta}$, and $\dfrac{d\psi_1}{d\theta}$ are given by Eqs. (5.2.11), (5.2.12), and (5.2.13), respectively.

5.4 Generation of Elliptical Gears by Shaper

DERIVATION OF SURFACE Σ_1 OF ELLIPTICAL GEAR GENERATED BY SHAPER. The derivation of the surface Σ_1 of a elliptical gear generated by a shaper is based on the following procedure:

(1) Two coordinate systems S_s and S_1 are considered rigidly connected to the shaper and the to-be-determined surface.
(2) An involute tooth surface $\mathbf{r}_s(u_s, \theta_s)$ is considered as given for a shaper with pitch radius ρ_s.

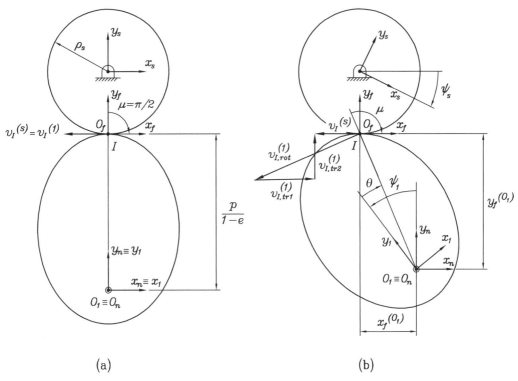

(a)   (b)

Figure 5.4.1. For determination of an elliptical gear generated by a shaper: (a) initial position; (b) current position.

(3) Coordinate system S_s performs rotation while coordinate system S_1 is rotated and translated. Motions of systems S_s and S_1 are defined by the following kinematic relation of their centrodes (see Fig. 5.4.1):

$$\mathbf{v}_I^{(s)} = \mathbf{v}_I^{(1)} \tag{5.4.1}$$

where

$$\mathbf{v}_I^{(1)} = \mathbf{v}_{I,rot}^{(1)} + \mathbf{v}_{I,tr1}^{(1)} + \mathbf{v}_{I,tr2}^{(1)} \tag{5.4.2}$$

(4) Coordinate system S_s is rotated an angle ψ_s that is related with polar angle θ by

$$\psi_s = \frac{s(\theta)}{\rho_s} \tag{5.4.3}$$

(5) Coordinate system S_n is translated with system S_1 on magnitudes $x_f^{(O_1)}$ and $y_f^{(O_1)}$ while system S_1 is rotated on the magnitude ψ_1. Such magnitudes may be represented as functions of polar angle θ:

$$x_f^{(O_1)} = -r(\theta)\cos\mu \tag{5.4.4}$$

$$y_f^{(O_1)} = r(\theta)\sin\mu \tag{5.4.5}$$

$$\psi_1 = \theta + \mu - \frac{\pi}{2} \tag{5.4.6}$$

(6) Elliptical gear-tooth surface Σ_1 is obtained by simultaneous consideration of matrix transformation

$$\mathbf{r}_1(u_s, \theta_s, \theta) = \mathbf{M}_{1s}(\theta)\mathbf{r}_s(u_s, \theta_s) \tag{5.4.7}$$

and equation of meshing

$$f(u_s, \theta_s, \theta) = \left(\frac{\partial \mathbf{r}_1}{\partial u_s} \times \frac{\partial \mathbf{r}_1}{\partial \theta_s}\right) \cdot \frac{\partial \mathbf{r}_1}{\partial \theta} = 0 \tag{5.4.8}$$

Derivation of Eq. (5.4.7) is performed as

$$\mathbf{r}_1(u_s, \theta_s, \theta) = \mathbf{M}_{1n}(\psi_1)\mathbf{M}_{nf}(x_f^{(O_1)}, y_f^{(O_1)})\mathbf{M}_{fs}(\psi_s)\mathbf{r}_s(u_s, \theta_s) \tag{5.4.9}$$

wherein

$$\mathbf{M}_{1n}(\psi_1) = \begin{bmatrix} \cos\psi_1 & \sin\psi_1 & 0 & 0 \\ -\sin\psi_1 & \cos\psi_1 & 0 & 0 \\ 0 & 0 & 1 & 0 \\ 0 & 0 & 0 & 1 \end{bmatrix} \tag{5.4.10}$$

$$\mathbf{M}_{nf}(x_f^{(O_1)}, y_f^{(O_1)}) = \begin{bmatrix} 1 & 0 & 0 & -x_f^{(O_1)} \\ 0 & 1 & 0 & -y_f^{(O_1)} \\ 0 & 0 & 1 & 0 \\ 0 & 0 & 0 & 1 \end{bmatrix} \tag{5.4.11}$$

$$\mathbf{M}_{fs}(\psi_s) = \begin{bmatrix} \cos\psi_s & \sin\psi_s & 0 & 0 \\ -\sin\psi_s & \cos\psi_s & 0 & \rho_s \\ 0 & 0 & 1 & 0 \\ 0 & 0 & 0 & 1 \end{bmatrix} \tag{5.4.12}$$

MATRIX DERIVATION OF EQUATION OF MESHING $f(u_s, \theta_s, \theta) = 0$. Matrix transformation given by Eq. (5.4.9) may be expressed by consideration of cartesian coordinates as

$$\boldsymbol{\rho}_1(u_s, \theta_s, \theta) = \mathbf{L}_{1n}\mathbf{L}_{nf}\mathbf{L}_{fs}\boldsymbol{\rho}_s(u_s, \theta_s) + \mathbf{R}$$

$$= \begin{bmatrix} \cos\psi_1 & \sin\psi_1 & 0 \\ -\sin\psi_1 & \cos\psi_1 & 0 \\ 0 & 0 & 1 \end{bmatrix} \begin{bmatrix} 1 & 0 & 0 \\ 0 & 1 & 0 \\ 0 & 0 & 1 \end{bmatrix} \begin{bmatrix} \cos\psi_s & \sin\psi_s & 0 \\ -\sin\psi_s & \cos\psi_s & 0 \\ 0 & 0 & 1 \end{bmatrix} \boldsymbol{\rho}_s(u_s, \theta_s) + \mathbf{R}$$

$$\tag{5.4.13}$$

where $\mathbf{R}$ is given by

$$\mathbf{R} = \begin{bmatrix} -x_f^{(O_1)} \cos \psi_1 - y_f^{(O_1)} \sin \psi_1 + \rho_s \sin \psi_1 \\ x_f^{(O_1)} \sin \psi_1 - y_f^{(O_1)} \cos \psi_1 + \rho_s \cos \psi_1 \\ 0 \end{bmatrix} \qquad (5.4.14)$$

Relative velocity $\mathbf{v}_1^{(s1)}$ is given as

$$\mathbf{v}_1^{(s1)} = \dot{\boldsymbol{\rho}}_1 = (\dot{\mathbf{L}}_{1n} \mathbf{L}_{nf} \mathbf{L}_{fs} + \mathbf{L}_{1n} \mathbf{L}_{nf} \dot{\mathbf{L}}_{fs}) \boldsymbol{\rho}_s + \dot{\mathbf{R}} \qquad (5.4.15)$$

wherein

$$\dot{\mathbf{L}}_{1n} = \begin{bmatrix} -\sin \psi_1 & \cos \psi_1 & 0 \\ -\cos \psi_1 & -\sin \psi_1 & 0 \\ 0 & 0 & 0 \end{bmatrix} \dot{\psi}_1$$

$$\dot{\mathbf{L}}_{fs} = \begin{bmatrix} -\sin \psi_s & \cos \psi_s & 0 \\ -\cos \psi_s & -\sin \psi_s & 0 \\ 0 & 0 & 0 \end{bmatrix} \dot{\psi}_s$$

$$\dot{\mathbf{R}} = \begin{bmatrix} -\dot{x}_f^{(O_1)} \cos \psi_1 - \dot{y}_f^{(O_1)} \sin \psi_1 + (x_f^{(O_1)} \sin \psi_1 - y_f^{(O_1)} \cos \psi_1 + \rho_s \cos \psi_1) \dot{\psi}_1 \\ \dot{x}_f^{(O_1)} \sin \psi_1 - \dot{y}_f^{(O_1)} \cos \psi_1 + (x_f^{(O_1)} \cos \psi_1 + y_f^{(O_1)} \sin \psi_1 - \rho_s \sin \psi_1) \dot{\psi}_1 \\ 0 \end{bmatrix}$$

Derivations of derivatives $\dot{\psi}_1$, $\dot{\psi}_s$, $\dot{x}_f^{(O_1)}$, and $\dot{y}_f^{(O_1)}$ are obtained as

$$\dot{\psi}_1 = \frac{d\psi_1}{d\theta} \frac{d\theta}{dt}$$

$$\dot{\psi}_s = \frac{d\psi_s}{d\theta} \frac{d\theta}{dt} = \frac{1}{\rho_s} \frac{ds}{d\theta} \frac{d\theta}{dt}$$

$$\dot{x}_f^{(O_1)} = \frac{dx_f^{(O_1)}}{d\theta} \frac{d\theta}{dt} = \frac{d(r \cos \mu)}{d\theta} \frac{d\theta}{dt} = \left(\frac{dr}{d\theta} \cos \mu - r \sin \mu \frac{d\mu}{d\theta} \right) \frac{d\theta}{dt}$$

$$\dot{y}_f^{(O_1)} = \frac{dy_f^{(O_1)}}{d\theta} \frac{d\theta}{dt}$$

Table 5.5.1. *Design parameters of planar elliptical gear.*

Module, m_n	2.0 mm
Pressure angle, α_c	20°
Eccentricity, e	0.5
Number of teeth, N_1	27
Half length of major axis, a	28.901 mm
Face width, w	15 mm

where

$$\frac{d\psi_1}{d\theta} = \frac{d}{d\theta}[\theta + \mu - \frac{\pi}{2}] = 1 + \frac{d\mu(\theta)}{d\theta}$$

$$\frac{ds(\theta)}{d\theta} = \frac{d}{d\theta}\int_0^\theta \frac{r(\theta)}{\sin\mu(\theta)}d\theta = \frac{r(\theta)}{\sin\mu(\theta)}$$

$$\frac{dr(\theta)}{d\theta} = \frac{d}{d\theta}\frac{p}{1 - e\cos\theta} = \frac{-pe\sin\theta}{(1 - e\cos\theta)^2}$$

$$\frac{d\mu(\theta)}{d\theta} = \frac{d}{d\theta}\arctan\frac{-1 + e\cos\theta}{e\sin\theta} = \frac{-e(e - \cos\theta)}{1 + e(e - 2\cos\theta)}$$

$$\frac{dy_f^{(O_1)}}{d\theta} = \frac{d}{d\theta}[-r(\theta)\sin\mu(\theta)] = -\frac{dr(\theta)}{d\theta}\sin\mu(\theta) - r(\theta)\cos\mu(\theta)\frac{d\mu(\theta)}{d\theta}$$

Then, the equation of meshing may be determined as

$$f_1^{(s1)}(u_s, \theta_s, \theta) = \mathbf{n}_1 \cdot \mathbf{v}_1^{(s1)} = \bar{\mathbf{L}}_{1s}\mathbf{n}_s \cdot \dot{\boldsymbol{\rho}}_1 = 0 \qquad (5.4.16)$$

or as

$$f_s^{(1s)}(u_s, \theta_s, \theta) = \mathbf{n}_s \cdot \mathbf{v}_s^{(1s)} = \mathbf{n}_s \cdot (-\mathbf{L}_{s1}\dot{\boldsymbol{\rho}}_1) = 0 \qquad (5.4.17)$$

5.5 Examples of Design of Planar and Helical Elliptical Gears

5.5.1 Planar Elliptical Gears

The design parameters applied are represented in Table 5.5.1. Parameter a has been obtained from the values m_n, e, and N_1 by application of (see Eq. (4.3.26))

$$a = \frac{N_1 m_n \pi}{\int_0^{2\pi}\sqrt{1 - e^2\sin^2\psi}\,d\psi}$$

The planar elliptical gear with parameters of Table 5.5.1 is shown in Fig. 5.5.1. The contact lines between the rack cutter and the gear (Fig. 5.5.2) are shown on the

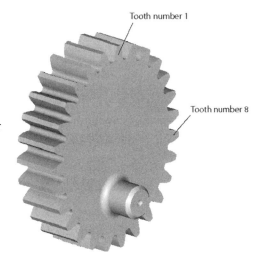

Figure 5.5.1. Illustration of planar elliptical gear with parameters shown in Table 5.5.1.

respective contacting surfaces: (a) for tooth number 1, and (b) tooth number 8 (see the notation of tooth numbers in Fig. 5.5.1).

A line of contact of generating and generated surfaces (the rack cutter and generated gear) is obtained as follows:

(i) The line of contact on the generating surface (on the rack cutter) is obtained considering simultaneously Eq. (5.2.14) of the rack cutter and equation of meshing (Eq. (5.2.25)) and considering parameter ψ_1 as constant. Parameter $\psi_1(\theta)$ is determined by Eqs. (5.2.6) and (5.2.7). Parameter β_c is equal to zero because a planar gear is generated.

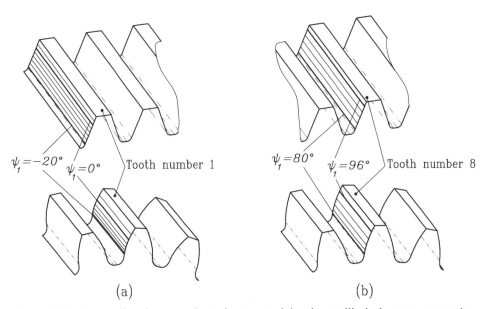

Figure 5.5.2. Contact lines between the rack cutter and the planar elliptical gear on contacting surfaces: (a) of tooth 1; (b) of tooth 8.

Table 5.5.2. *Design parameters of helical elliptical gear.*

Module, m_n	2.0 mm
Pressure angle, α_c	20°
Eccentricity, e	0.5
Helix angle, β_c	20° (left-hand)
Number of teeth, N_1	27
Half length of major axis, a	30.756 mm
Face width, w	15 mm

(ii) The line of contact on the generated surface of the elliptical gear is obtained considering simultaneously Eq. (5.2.18) (or (5.2.20)) of the family of generating surfaces and equation of meshing (Eq. (5.2.25)) and considering again ψ_1 as constant.

5.5.2 Helical Elliptical Gears

The designed parameters for the discussed example are represented in Table 5.5.2. Parameter a has been obtained from the values m_n, β_c, e, and N_1 by application of (see Eq. (4.3.26))

$$a = \frac{N_1 m \pi}{\displaystyle\int_0^{2\pi} \sqrt{1 - e^2 \sin^2 \psi}\, d\psi}$$

wherein $m = m_n / \cos \beta_c$.

The helical elliptical gear with design parameters represented in Table 5.5.2 is shown in Fig. 5.5.3. The contact lines between the rack cutter and the gear (Figs. 5.5.4(a) and (b)) are shown on the respective contacting surfaces: (a) on the rack

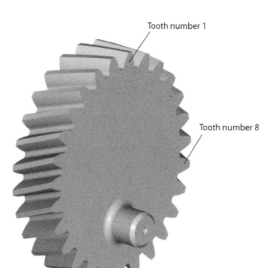

Tooth number 1

Tooth number 8

Figure 5.5.3. Illustration of helical elliptical gear with parameters shown in Table 5.5.2.

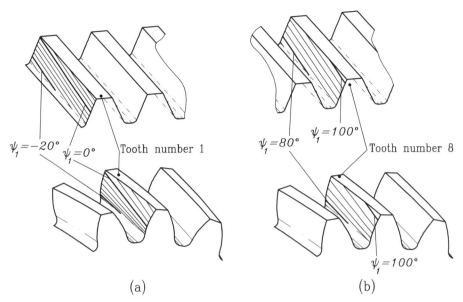

Figure 5.5.4. Contact lines between the rack cutter and the helical elliptical gear on contacting surfaces: (a) of tooth 1; (b) of tooth 8.

cutter, the contact lines are parallel straight lines; (b) on the helical elliptical gear, the contact lines are straight lines as well. The contact lines shown in Figs. 5.5.4(a) and (b) have been determined for tooth numbers 1 and 8 (Fig. 5.5.3). The contact lines discussed previously are determined by application of Eq. (5.2.20) of family of generating rack-cutter surfaces and equation of meshing (Eq. (5.2.25)), but by considering ψ_1 as constant.

6 Design of Gear Drives Formed by Eccentric Circular Gear and Conjugated Noncircular Gear

6.1 Introduction

The advantage of a gear drive formed by an eccentric circular gear and its conjugated noncircular gear (called for the purpose of abbreviation *eccentric gear drive*) is the simplification of the generation of the driving gear designed as a circular gear with displaced center of rotation. Figure 6.1.1 shows an eccentric involute pinion and the conjugated noncircular gear. The noncircular gear of the eccentric gear drive may be designed and generated with straight or helical teeth.

The eccentric gear drive is competitive to the one formed by elliptical gears (see Chapter 5). In this chapter, the geometry and design of eccentric gear drives will be described and methods of generation of the eccentric gear and its conjugated noncircular gear by a hob (in case of having convex centrodes) and by a shaper (in case of having convex-concave centrodes) will be presented.

6.2 Centrodes of Eccentric Gear Drive

6.2.1 Equations of Mating Centrodes

Figures 6.2.1 shows centrodes σ_1 and σ_2 in initial and current positions. Centrode σ_1 is an eccentric circle of radius r_{p1} and is represented in polar form as (Fig. 6.2.1(b))

$$r_1(\theta_1) = \left(r_{p1}^2 - e_1^2 \sin^2 \theta_1\right)^{\frac{1}{2}} - e_1 \cos \theta_1 \qquad (6.2.1)$$

where parameter θ_1 determines the position of $r_1(\theta_1)$ with respect to polar axis $\overline{O_1 A_1}$ (Fig. 6.2.1(b)).

Henceforth, we will use representation of a centrode in terms of derivative function $m_{12}(\phi_1)$ (see Section 2.4), where $\phi_i = \theta_i$ $(i = 1, 2)$ is the angle of rotation of the centrodes; parameter θ_i determines location of $r_i(\theta_i)$ with respect to polar axis $\overline{O_i A_i}$. For σ_1, we have

$$r_1(\phi_1) = \frac{E}{1 + m_{12}(\phi_1)} \qquad (6.2.2)$$

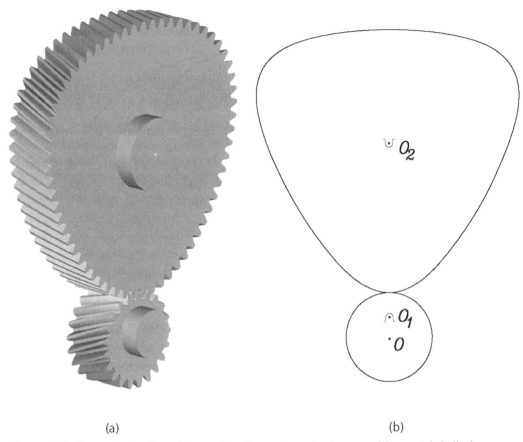

(a)   (b)

Figure 6.1.1. Eccentric involute pinion and conjugated noncircular gear: (a) eccentric helical drive, (b) eccentric planar drive. O_1 and O_2 are the centers of rotation of pinion and gear, respectively; O is the geometric center of the pinion.

Centrode σ_2 is represented by

$$r_2(\phi_1) = E \frac{m_{12}(\phi_1)}{1 + m_{12}(\phi_1)} \qquad (6.2.3)$$

$$\phi_2(\phi_1) = \int_0^{\phi_1} \frac{d\phi_1}{m_{12}(\phi_1)} \qquad (6.2.4)$$

Here, the derivative function is

$$m_{12}(\phi_1) = \frac{E - r_1(\phi_1)}{r_1(\phi_1)} = \frac{c_1}{\left(1 - \varepsilon_1^2 \sin^2 \phi_1\right)^{\frac{1}{2}} - \varepsilon_1 \cos \phi_1} - 1 \qquad (6.2.5)$$

where $c_1 = E/r_{p1}$ and $\varepsilon_1 = e_1/r_{p1}$. Function $\phi_2(\phi_1)$ is the transmission function, and its determination requires numerical integration.

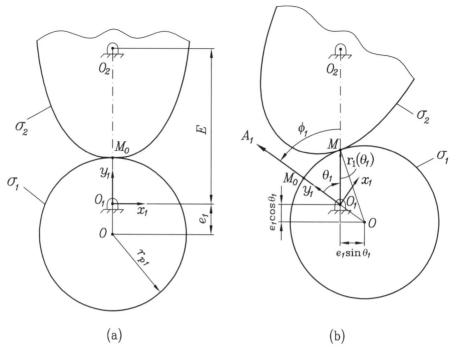

(a)   (b)

Figure 6.2.1. For derivation of centrodes: (a) initial position; (b) current position.

Centrode σ_1 is already a closed-form curve (a circle of radius r_{p1}). Centrode σ_2 will be a closed-form curve by observation of

$$\phi_2(\phi_1) = \frac{2\pi}{n} = \int_0^{2\pi} \frac{d\phi_1}{m_{12}(\phi_1)} \qquad (6.2.6)$$

wherein n is the number of revolutions that the eccentric gear 1 performs for one revolution of the noncircular gear.

Derivation of centrodes σ_1 and σ_2 enables us to design as well a cam mechanism with a constant center distance between the follower as an eccentric circle and a cam for generation of functions. The cam is determined as the centrode of the noncircular gear of the eccentric drive.

6.2.2 Curvature of Centrode σ_2 and Applications

BASIC DERIVATIONS. Knowledge of curvature of centrode σ_2 is necessary for: (i) choosing the method of generation of the gears by a hob or by a shaper, and (ii) for avoidance of undercutting of tooth profiles.

Considering that a centrode is represented by a polar curve $r(\theta)$, its curvature radius may be determined (see Section 2.8) as

$$\rho(\theta) = \frac{\left[r(\theta)^2 + \left(\dfrac{dr}{d\theta} \right)^2 \right]^{\frac{3}{2}}}{r(\theta)^2 + 2 \left(\dfrac{dr}{d\theta} \right)^2 - r(\theta) \dfrac{d^2 r}{d\theta^2}} \qquad (6.2.7)$$

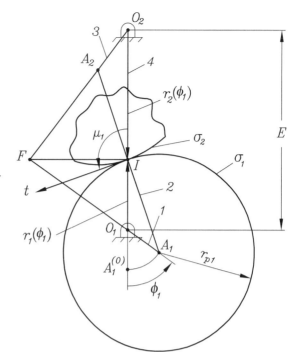

Figure 6.2.2. For derivation of curvature radius $\rho_2(\phi_1)$ of centrode σ_2.

The condition of convexity of the polar curve may be represented as

$$\kappa = \frac{1}{\rho} = r(\theta)^2 + 2\left(\frac{dr}{d\theta}\right)^2 - r(\theta)\frac{d^2r}{d\theta^2} \geq 0 \qquad (6.2.8)$$

or by an alternative inequality

$$\sin\mu\left[r(\theta)(2 - \sin^2\mu) - \sin^2\mu\frac{d^2r}{d\theta^2}\right] \geq 0 \qquad (6.2.9)$$

Here, angle $\mu = \arctan\left(\dfrac{r(\theta)}{dr/d\theta}\right)$ is formed by position vector $r(\theta)$ and tangent t to the polar curve (see Section 2.5); κ is the curvature of the polar curve.

Taken into account that the curvature of centrode σ_1 is known, determination of κ_2 can be obtained by using Euler-Savary equation (Hartenberg & Denavit, 1964), which relates curvatures κ_1 and κ_2 of centrodes σ_1 and σ_2 of the eccentric gear drive (Fig. 6.2.2) as

$$\frac{1}{\rho_1} + \frac{1}{\rho_2} = \left[\frac{1}{r_1(\phi_1)} + \frac{1}{r_2(\phi_1)}\right]\sin\mu_1 \qquad (6.2.10)$$

Here,

$$\frac{1}{\rho_1} = \frac{1}{r_{p1}} = \kappa_1, \qquad \frac{1}{\rho_2} = \kappa_2 \qquad (6.2.11)$$

where $\rho_2 = |\overline{I\,A_2}|$ (Fig. 6.2.2), and

$$r_2(\phi_1) = E - r_1(\phi_1) \qquad (6.2.12)$$

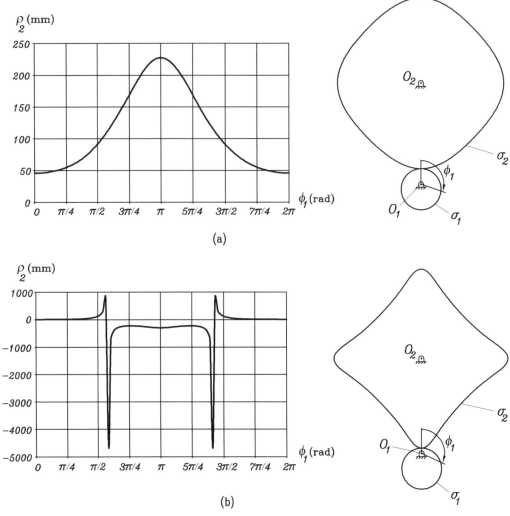

Figure 6.2.3. Illustration of function $\rho_2(\phi_1)$ for eccentric gear drives with design parameters $r_{p1} = 22.35$ mm, $n = 4$, and (a) $\varepsilon = 0.2$, (b) $\varepsilon = 0.7$.

Equation (6.2.10) yields

$$\kappa_2 = \left[\frac{1}{r_1(\phi_1)} + \frac{1}{E - r_1(\phi_1)} \right] \sin \mu_1 - \kappa_1 \qquad (6.2.13)$$

Convexity of centrode σ_2 is guaranteed if

$$\left[\frac{1}{r_1(\phi_1)} + \frac{1}{E - r_1(\phi_1)} \right] \sin \mu_1 - \kappa_1 \geq 0 \qquad (6.2.14)$$

Figure 6.2.3 illustrates function $\rho_2(\phi_1)$ for gear drives with: (a) a convex centrode σ_2; (b) a convex-concave centrode σ_2. The location and orientation of curvature radius $\rho_2(\phi_1)$ is visualized by application of a four-bar linkage (Fig. 6.2.2): (a) link 1 as $\overline{O_1 A_1}$, (b) link 2 as $\overline{A_1 A_2}$, (c) link 3 as $\overline{O_2 A_2}$, and (d) link 4 as $\overline{O_1 O_2}$, where $|\overline{A_1 A_2}| = \rho_1 + \rho_2$.

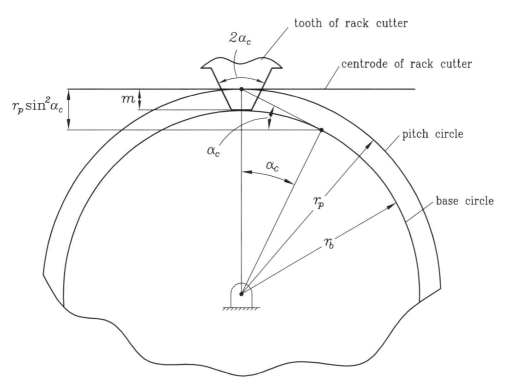

Figure 6.2.4. Illustration of avoidance of undercutting of a spur involute pinion as the condition $m < r_p \sin^2 \alpha_c$.

It is known from kinematics of a four-bar linkage that point F (of intersection of extended links 1 and 3) is the instantaneous center of rotation of link 2 with respect to link 4. Vector $\overline{IF}$ is perpendicular to $\overline{O_1 O_2}$, and its magnitude is determined as

$$|\overline{IF}| = \frac{r_1(\phi_1) - \rho_1 \sin \mu_1}{\rho_1 r_1(\phi_1)} = \frac{|\rho_2 \sin \mu_1 - r_2(\phi_1)|}{\rho_2 r_2(\phi_1)} \tag{6.2.15}$$

AVOIDANCE OF UNDERCUTTING. Drawings of Fig. 6.2.4 show generation of an involute spur gear with a rack cutter of profile angle α_c. The centrode of the rack cutter is its middle line. The centrode of the gear is the pitch circle of radius r_p. Undercutting of the involute gear is avoided by the installment of the rack cutter wherein

$$\frac{m}{\sin^2 \alpha_c} \leq r_p \tag{6.2.16}$$

Avoidance of undercutting of the noncircular gear of eccentric gear drive is based on the following considerations:

(i) Gear 2 of the drive has various tooth profiles, but they may be represented (approximately) as tooth profiles of respective circular gears with curvatures of radii $\rho_A, \rho_B, \ldots, \rho_K$ (Fig. 6.2.5).

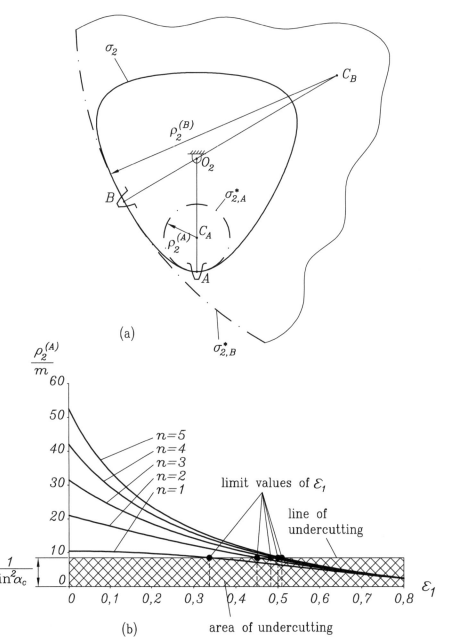

Figure 6.2.5. Illustration of: (a) representation of tooth profiles of noncircular gear 2 ($\sigma_{2,A}^*$ is the centrode of substituting circular gear at A and $\sigma_{2,B}^*$ is the centrode of substituting circular gear at B); (b) line of undercutting for the case wherein $\alpha_c = 20°$.

(ii) Radius $\rho_2^{(A)}$ of curvature of centrode σ_2 for point A corresponds to profiles of the tooth denoted as Σ_A. Similarly, profiles of tooth Σ_B are represented as those of spur gears of radius $\rho_2^{(B)}$, and so on (Fig. 6.2.5).

(iii) Undercutting occurs for a tooth with the smallest radius ρ_2 of the substituting circular gear.

(iv) Figure 6.2.5(a) shows representation of tooth profiles of gear 2 with centrode σ_2 designed for a gear drive with $n = 3$. Here, n is the number of revolutions of eccentric gear 1 that is performed for one revolution of driven gear 2 with centrode σ_2.

(v) The idea of application of substituting circular gears may be applied for avoidance of undercutting for all eccentric gear drives with centrodes σ_1 and σ_2. For this purpose, it is necessary to obtain functions $\rho_2^{(A)}(\varepsilon, n)$ (Fig. 6.2.5(b)) that represent the minimal curvature radius of ρ_2. Point A is the point of centrode σ_2 (Fig. 6.2.5(a)) where $\rho_2 = \rho_{2,min}$, and $\varepsilon = e/r_p$ (see Eq. (6.2.5)).

(vi) Avoidance of undercutting for all eccentric gear drives is obtained by choosing such a module for which functions $\frac{\rho_2^{(A)}}{m}$ will be out of the square with the height $1/\sin^2 \alpha_c$ (Fig. 6.2.5(b)).

6.3 Generation of the Noncircular Gear by Shaper and Hob

In this section, the algorithms that relate the motions of the generating tool (shaper, hob) and the noncircular gear of the drive being generated will be formulated. Such relations are represented by nonlinear equations and are the basis of computerized generation of the noncircular gear.

The eccentric gear of the drive may be generated as a conventional involute gear, with the observation by manufacturing and assembly of the location of the geometric center with respect to the center of rotation. In addition, for the purpose of localization of the bearing contact it is necessary to provide double-crowning to the surfaces of the teeth of the eccentric gear.

There is the possibility of generation of the noncircular gear by a shaper that is identical to the eccentric involute gear of the drive. This allows manufacturing the noncircular gear observing a constant center distance between the shaper and noncircular gear being generated. However, the identity of the shaper and the eccentric involute gear has an unfavorable limitation and has not been applied by the authors.

6.3.1 Generation of Noncircular Gear by a Noneccentric Shaper

DERIVATION OF SURFACE Σ_2 OF NONCIRCULAR GEAR GENERATED BY SHAPER. We recall that the shaper is not identical to the involute eccentric gear of the drive. The derivation is based on the following procedure:

(1) Two coordinate systems S_s and S_2 are considered rigidly connected to the shaper and the to-be-determined surface.

(2) An involute tooth surface $\mathbf{r}_s(u_s, v_s)$ is considered as given for a shaper with pitch radius ρ_s, which may differ from ρ_1.

(3) Coordinate system S_s is rotated while coordinate system S_2 is rotated and translated. Motions of systems S_s and S_2 are defined by kinematic relation of their centrodes (see Fig. 6.3.1):

$$\mathbf{v}_I^{(s)} = \mathbf{v}_I^{(2)} \tag{6.3.1}$$

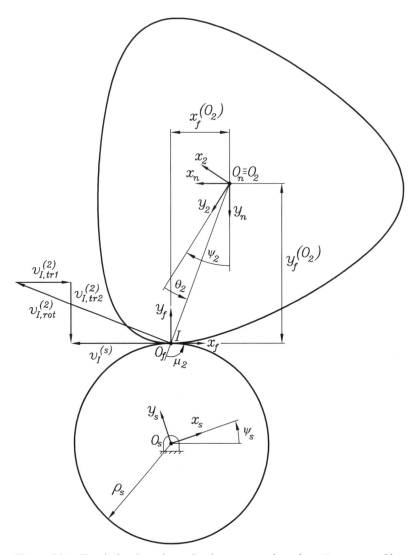

Figure 6.3.1. For derivation of noncircular gear tooth surface Σ_2 generated by a regular involute shaper.

where

$$\mathbf{v}_I^{(2)} = \mathbf{v}_{I,rot}^{(2)} + \mathbf{v}_{I,tr1}^{(2)} + \mathbf{v}_{I,tr2}^{(2)} \tag{6.3.2}$$

(4) Coordinate system S_s is rotated on ψ_s that is related with polar angle θ_2 by

$$\psi_s = \frac{s(\theta_2)}{\rho_s} \tag{6.3.3}$$

wherein $s(\theta_2)$ is a function of the polar angle θ_1 of the eccentric gear,

$$s(\theta_2) = \rho_1(\theta_1 - \arcsin(\varepsilon_1 \sin \theta_1)) \tag{6.3.4}$$

(5) Coordinate system S_n is translated with system S_2 on magnitudes $x_f^{(O_2)}$ and $y_f^{(O_2)}$ while system S_2 is rotated on the magnitude ψ_2. Such magnitudes may be represented as functions of polar angle θ_2:

$$x_f^{(O_2)} = -r(\theta_2)\cos\mu_2 \tag{6.3.5}$$

$$y_f^{(O_2)} = r(\theta_2)\sin\mu_2 \tag{6.3.6}$$

$$\psi_2 = \theta_2 + \mu_2 - \frac{\pi}{2} \tag{6.3.7}$$

Functions $r(\theta_2)$, θ_2, and μ_2 may be expressed as functions of polar angle θ_1 as

$$r(\theta_2) = E - r(\theta_1) = E - \rho_1\left(\sqrt{1 - \varepsilon_1^2 \sin^2\theta_1} - \varepsilon_1\cos\theta_1\right) \tag{6.3.8}$$

$$\theta_2 = \int_0^{\theta_1}\left(\frac{c_1}{c_1 - \sqrt{1 - \varepsilon_1^2\sin^2\theta_1} + \varepsilon_1\cos\theta_1} - 1\right)d\theta_1 \tag{6.3.9}$$

$$\mu_2(\theta_1) = \arctan\left(\frac{\sqrt{1 - \varepsilon_1^2\sin^2\theta_1}}{-\varepsilon_1\sin\theta_1}\right), \tag{6.3.10}$$

(6) Noncircular gear-tooth surface Σ_2 is obtained by simultaneous consideration of matrix transformation

$$\mathbf{r}_2(u_s, v_s, \theta_2) = \mathbf{M}_{2s}(\theta_2)\mathbf{r}_s(u_s, v_s) \tag{6.3.11}$$

and equation of meshing

$$f(u_s, v_s, \theta_2) = \left(\frac{\partial\mathbf{r}_2}{\partial u_s} \times \frac{\partial\mathbf{r}_2}{\partial v_s}\right)\cdot\frac{\partial\mathbf{r}_2}{\partial\theta_2} = 0 \tag{6.3.12}$$

Derivation of Eq. (6.3.11) is performed as

$$\mathbf{r}_2(u_s, v_s, \theta_2) = \mathbf{M}_{2n}(\psi_2)\mathbf{M}_{nf}(x_f^{(O_2)}, y_f^{(O_2)})\mathbf{M}_{fs}(\psi_s)\mathbf{r}_s(u_s, v_s) \tag{6.3.13}$$

Here,

$$\mathbf{M}_{2n} = \begin{bmatrix} \cos\psi_2 & -\sin\psi_2 & 0 & 0 \\ \sin\psi_2 & \cos\psi_2 & 0 & 0 \\ 0 & 0 & 1 & 0 \\ 0 & 0 & 0 & 1 \end{bmatrix}, \quad \mathbf{M}_{nf} = \begin{bmatrix} 1 & 0 & 0 & -x_f^{(O_2)} \\ 0 & 1 & 0 & -y_f^{(O_2)} \\ 0 & 0 & 1 & 0 \\ 0 & 0 & 0 & 1 \end{bmatrix}$$

$$\mathbf{M}_{fs} = \begin{bmatrix} \cos\psi_s & -\sin\psi_s & 0 & 0 \\ \sin\psi_s & \cos\psi_s & 0 & -\rho_s \\ 0 & 0 & 1 & 0 \\ 0 & 0 & 0 & 1 \end{bmatrix}$$

MATRIX DERIVATION OF EQUATION OF MESHING $f(u_s, v_s, \theta_2) = 0$. Matrix transformation Eq. (6.3.13) may be expressed by consideration of Cartesian coordinates as

$$\rho_2(u_s, v_s, \theta_2) = \mathbf{L}_{2n}\mathbf{L}_{nf}\mathbf{L}_{fs}\rho_s(u_s, v_s) + \mathbf{R} \tag{6.3.14}$$

The 3×3 matrices $\mathbf{L}$ are obtained from 4×4 matrices $\mathbf{M}$ by eliminating the last row and last column; $\mathbf{R}$ is given by

$$\mathbf{R} = \begin{bmatrix} -x_f^{(O_2)}\cos\psi_2 + y_f^{(O_2)}\sin\psi_2 + \rho_s\sin\psi_2 \\ -x_f^{(O_2)}\sin\psi_2 - y_f^{(O_2)}\cos\psi_2 - \rho_s\cos\psi_2 \\ 0 \end{bmatrix} \tag{6.3.15}$$

Relative velocity $\mathbf{v}_2^{(s2)}$ is given as

$$\mathbf{v}_2^{(s2)} = \dot{\boldsymbol{\rho}}_2 = (\dot{\mathbf{L}}_{2n}\mathbf{L}_{nf}\mathbf{L}_{fs} + \mathbf{L}_{2n}\mathbf{L}_{nf}\dot{\mathbf{L}}_{fs})\boldsymbol{\rho}_s + \dot{\mathbf{R}} \tag{6.3.16}$$

Derivations of derivatives $\dot{\psi}_2$, $\dot{\psi}_s$, $\dot{x}_f^{(O_2)}$, and $\dot{y}_f^{(O_2)}$ are obtained as

$$\dot{\psi}_2 = \left(\frac{d\theta_2}{d\theta_1} + \frac{d\mu_2}{d\theta_1}\right)\frac{d\theta_1}{dt}$$

$$\dot{\psi}_s = \frac{1}{\rho_s}\frac{ds(\theta_2)}{d\theta_1}\frac{d\theta_1}{dt}$$

$$\dot{x}_f^{(O_2)} = \frac{dx_f^{(O_2)}}{d\theta_1}\frac{d\theta_1}{dt}$$

$$\dot{y}_f^{(O_2)} = \frac{dy_f^{(O_2)}}{d\theta_1}\frac{d\theta_1}{dt}$$

Derivatives $\dfrac{d\theta_2}{d\theta_1}, \dfrac{d\mu_2}{d\theta_1}, \dfrac{ds(\theta_2)}{d\theta_1}, \dfrac{dx_f^{(O_2)}}{d\theta_1}$, and $\dfrac{dy_f^{(O_2)}}{d\theta_1}$ are given by:

$$\frac{d\theta_2}{d\theta_1} = \frac{c_1}{c_1 - \sqrt{1 - \varepsilon_1^2\sin^2\theta_1} + \varepsilon_1\cos\theta_1} - 1 \tag{6.3.17}$$

$$\frac{d\mu_2}{d\theta_1} = \varepsilon_1^2\sin^2\theta_1 \tag{6.3.18}$$

$$\frac{ds(\theta_2)}{d\theta_1} = \rho_1\left(1 - \frac{1}{\sqrt{1 - \varepsilon_1^2\sin^2\theta_1}}\right) \tag{6.3.19}$$

$$\frac{dx_f^{(O_2)}}{d\theta_1} = -\rho_1\left(\frac{\varepsilon_1^2\sin\theta_1\cos\theta_1}{\sqrt{1 - \varepsilon_1^2\sin^2\theta_1}} - \varepsilon_1\sin\theta_1\right)\cos\mu_2$$
$$+ r(\theta_2)\sin\mu_2\varepsilon_1^2\sin^2\theta_1 \tag{6.3.20}$$

$$\frac{dy_f^{(O_2)}}{d\theta_1} = \rho_1\left(\frac{\varepsilon_1^2\sin\theta_1\cos\theta_1}{\sqrt{1 - \varepsilon_1^2\sin^2\theta_1}} - \varepsilon_1\sin\theta_1\right)\sin\mu_2$$
$$+ r(\theta_2)\cos\mu_2\varepsilon_1^2\sin^2\theta_1 \tag{6.3.21}$$

Then, the equation of meshing may be determined as

$$f_2^{(s2)}(u_s, v_s, \theta_1) = \mathbf{n}_2 \cdot \mathbf{v}_2^{(s2)} = \mathbf{L}_{2s}\mathbf{n_s} \cdot \dot{\boldsymbol{\rho}}_2 = 0 \tag{6.3.22}$$

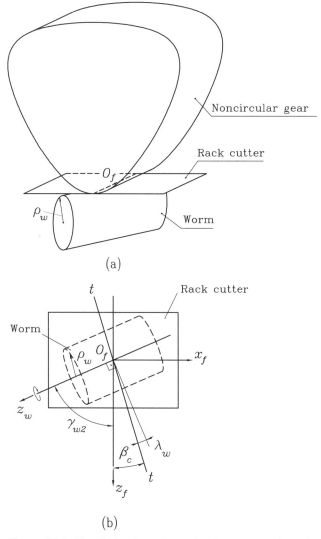

Figure 6.3.2. For derivation of noncircular gear-tooth surface Σ_2 generated by a hob: (a) installation of noncircular gear, worm, and imaginary rack cutter; (b) installation of worm respect to imaginary rack cutter.

or as

$$f_s^{(2s)}(u_s, v_s, \theta_1) = \mathbf{n_s} \cdot \mathbf{v}_s^{(2s)} = \mathbf{n_s} \cdot (-\mathbf{L}_{s2}\dot{\boldsymbol{\rho}}_2) = 0 \qquad (6.3.23)$$

6.3.2 Generation of the Noncircular Gear by a Hob

INSTALLATION OF HOB. Generation of noncircular gear-tooth surface Σ_2 by a worm-thread surface Σ_w is being considered. For the purpose of simplicity, an imaginary rack-cutter tooth surface Σ_c is considered in continuous tangency with surfaces Σ_2 and Σ_w.

Figure 6.3.2(a) shows worm, rack cutter, and noncircular gear at their initial positions. Point O_f is the common point of tangency of three surfaces Σ_2, Σ_w, and Σ_c.

Worm shaft and noncircular gear shaft form crossing angle γ_{w2} (see Fig. 6.3.2(b)). Angle γ_{w2} is given as

$$\gamma_{w2} = \frac{\pi}{2} - \beta_c - \lambda_w \qquad (6.3.24)$$

wherein β_c is the helix angle of the skew rack cutter and λ_w is the lead angle of the worm.

For the purpose of simplicity, worm-thread surface may be considered as a surface with two independent parameters (h_w, v_w).

GENERATION OF SURFACE Σ_2 OF NONCIRCULAR GEAR BY WORM. The derivation of Σ_2 by a worm-thread surface Σ_w is based on the following procedure (see Fig. 6.3.3):

(i) Two coordinate systems S_w and S_2 are considered rigidly connected to the worm-thread surface and the to-be-determined gear tooth surface. A fixed reference system S_f is considered for definition of motions of systems S_w and S_2.
(ii) Worm-thread surface Σ_w is considered as given by vector $\mathbf{R}_w(h_w, v_w)$.
(iii) Two sets of motions are provided to the worm:
    (a) Rotation ϕ_w about axis z_w of the worm.
    (b) Translation s_w along axis z_f that is parallel to the axis of the gear.
    Coordinate system S_s is a movable coordinate system that is translated with system S_w.
(iv) Rotation and translation of the worm are accompanied by rotation and translation of the elliptical gear as follows:
    (a) Rotation ψ_2 about axis z_2 of the gear.
    (b) Translation defined by position $y_f^{(O_2)}$ along axis y_f.
    Magnitudes ψ_2 and $y_f^{(O_2)}$ may be determined as functions of polar angle θ_2 of the noncircular gear as

$$y_f^{(O_2)}(\theta_2) = r(\theta_2) \sin \mu_2(\theta_2) \qquad (6.3.25)$$

$$\psi_2(\theta_2) = \theta_2 + \mu_2(\theta_2) - \mu_{20}, \qquad \mu_{20} = \frac{\pi}{2} \qquad (6.3.26)$$

Because $r(\theta_2)$, θ_2, and μ_2 are related with polar angle θ_1 of the eccentric gear by functions

$$r(\theta_2) = E_{12} - r(\theta_1) = E_{12} - \rho_1\left(\sqrt{1 - \varepsilon_1^2 \sin^2 \theta_1} - \varepsilon_1 \cos \theta_1\right) \quad (6.3.27)$$

$$\theta_2 = \int_0^{\theta_1}\left(\frac{c_1}{c_1 - \sqrt{1 - \varepsilon_1^2 \sin^2 \theta_1} + \varepsilon_1 \cos \theta_1} - 1\right) d\theta_1 \qquad (6.3.28)$$

$$\mu_2(\theta_1) = \arctan\left(\frac{\sqrt{1 - \varepsilon_1^2 \sin^2 \theta_1}}{-\varepsilon_1 \sin \theta_1}\right), \qquad (6.3.29)$$

a new function $g(\phi_w, s_w, \theta_1) = 0$ must be determined to relate ϕ_w and s_w with polar angle θ_1.

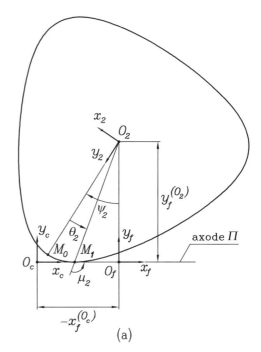

Figure 6.3.3. Applied coordinate systems for derivation of noncircular gear-tooth surface Σ_2 generated by a hob: (a) systems S_2 and S_c of noncircular gear and imaginary rack cutter; (b) system S_w of worm and auxiliary system S_s.

(a)

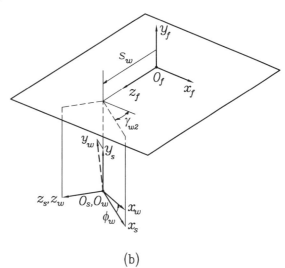

(b)

(v) Coordinate transformation between systems S_w and S_2 determines the family of worm-thread surfaces in system S_2 as

$$\mathbf{r}_2(h_w, v_w, \phi_w, s_w) = \mathbf{M}_{2w}(\phi_w, s_w)\mathbf{R}_w(h_w, v_w) \qquad (6.3.30)$$

Here, ϕ_w and s_w are independent generalized parameters of motion, which means that the generation is a double-enveloping process; $\mathbf{M}_{2w}$ is a 4×4 matrix that describes coordinate transformation from system S_w to system S_2.

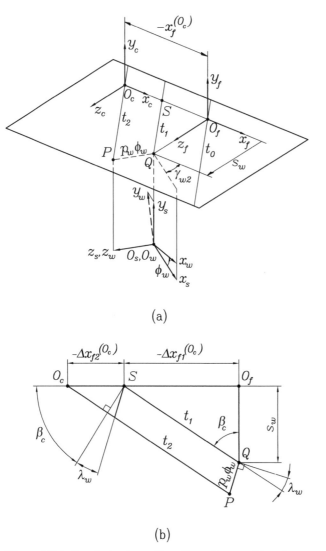

(a)

(b)

Figure 6.3.4. For derivation of function $g(\phi_w, s_w, \theta_1) = 0$.

(vi) Equations of meshing

$$f_1^{(w2)}(h_w, v_w, \phi_w, s_w) = 0 \qquad (6.3.31)$$

$$f_2^{(w2)}(h_w, v_w, \phi_w, s_w) = 0 \qquad (6.3.32)$$

relate parameters (h_w, v_w, ϕ_w, s_w). Simultaneous consideration of Eqs. (6.3.30), (6.3.31), and (6.3.32) determines surface Σ_2 of noncircular gear.

DERIVATION OF FUNCTION $g(\phi_w, s_w, \theta_1) = 0$. The derivation is performed as follows (see Fig. 6.3.4):

(i) An imaginary rack cutter is considered in simultaneous meshing with the non-circular gear and the worm. At the initial position, system S_c coincides with system S_f, and common tangent line $t-t$ between the three surfaces, Σ_w, Σ_2, and Σ_c, is at position t_0.

(ii) Due to rotation and translation of the worm on ϕ_w and s_w, the common tangent $t-t$ will take position t_2. The location of system S_c in S_f is determined by $x_f^{(O_c)}$.

(iii) Displacement of system S_c may be obtained as the sum of independent displacements $\Delta x_{f1}^{(O_c)}$ and $\Delta x_{f2}^{(O_c)}$:

$$\Delta x_f^{(O_c)} = \Delta x_{f1}^{(O_c)} + \Delta x_{f2}^{(O_c)} \tag{6.3.33}$$

Displacement $\Delta x_{f1}^{(O_c)} = \overline{O_f S}$ is caused by translation s_w and is defined by positions t_0 and t_1. Displacement $\Delta x_{f2}^{(O_c)} = \overline{S O_c}$ is a result of rotation ϕ_w and is defined by positions t_1 and t_2.

(iv) Illustrations of Fig. 6.3.4(b) yield

$$-\Delta x_{f1}^{(O_c)} = \overline{O_f S} = \tan \beta_c s_w \tag{6.3.34}$$

$$-\Delta x_{f2}^{(O_c)} = \overline{S O_C} = \frac{p_w \cos \lambda_w}{\cos \beta_c} \phi_w \tag{6.3.35}$$

where p_w is the pitch of the worm.

(v) Because $x_f^{(O_c)}$ depends on polar angle θ_1 as

$$x_f^{(O_c)} = -s(\theta_2) + r(\theta_2) \cos \mu_2(\theta_2) \tag{6.3.36}$$

wherein

$$s(\theta_2) = \rho_1 \left(\theta_1 - \arcsin(\varepsilon_1 \sin \theta_1) \right) \tag{6.3.37}$$

$$r(\theta_2) = E_{12} - \rho_1 \left(\sqrt{1 - \varepsilon_1^2 \sin^2 \theta_1} - \varepsilon_1 \cos \theta_1 \right) \tag{6.3.38}$$

$$\mu_2(\theta_2) = \arctan \left(\frac{\sqrt{1 - \varepsilon_1^2 \sin^2 \theta_1}}{-\varepsilon_1 \sin \theta_1} \right) \tag{6.3.39}$$

function $g(\phi_w, s_w, \theta_1) = 0$ is obtained finally as

$$g(\phi_w, s_w, \theta_1) = x_f^{(O_c)}(\theta_1) + \tan \beta_c s_w + \frac{p_w \cos \lambda_w}{\cos \beta_c} \phi_w = 0 \tag{6.3.40}$$

COORDINATE TRANSFORMATION IN TRANSITION FROM COORDINATE SYSTEM S_w TO S_2. Derivation of Eq. (6.3.30) is performed as

$$\mathbf{r}_2(h_w, v_w, \phi_w, s_w) = \mathbf{M}_{2f}(\phi_w, s_w)\mathbf{M}_{fs}(s_w)\mathbf{M}_{sw}(\phi_w)\mathbf{R}_w(h_w, v_w) \tag{6.3.41}$$

Here,

$$\mathbf{M}_{2f} = \begin{bmatrix} -\cos \psi_2 & \sin \psi_2 & 0 & y_f^{(O_2)} \sin \psi_2 \\ -\sin \psi_2 & -\cos \psi_2 & 0 & -y_f^{(O_2)} \cos \psi_2 \\ 0 & 0 & 1 & 0 \\ 0 & 0 & 0 & 1 \end{bmatrix},$$

$$\mathbf{M}_{fs} = \begin{bmatrix} \cos \gamma_{w2} & 0 & -\sin \gamma_{w2} & 0 \\ 0 & 1 & 0 & -\rho_w \\ \sin \gamma_{wg} & 0 & \cos \gamma_{wg} & s_w \\ 0 & 0 & 0 & 1 \end{bmatrix}$$

$$\mathbf{M}_{sw} = \begin{bmatrix} \cos \phi_w & -\sin \phi_w & 0 & 0 \\ \sin \phi_w & \cos \phi_w & 0 & 0 \\ 0 & 0 & 1 & 0 \\ 0 & 0 & 0 & 1 \end{bmatrix}$$

MATRIX DERIVATION OF EQUATIONS OF MESHING. Matrix derivation of equations of meshing $f_1^{(w2)}(h_w, v_w, \phi_w, s_w) = 0$ and $f_2^{(w2)}(h_w, v_w, \phi_w, s_w) = 0$ is applied as follows:

(i) Position vector $\mathbf{r}_2(h_w, v_w, \phi_w, s_w)$ is represented in homogeneous coordinates by

$$\mathbf{r}_2(h_w, v_w, \phi_w, s_w) = \mathbf{M}_{2w}(\phi_w, s_w)\mathbf{R}_w(h_w, v_w)$$

$$= \begin{bmatrix} a_{11} & a_{12} & a_{13} & \sin \psi_2(y_f^{(O_2)} - \rho_w) \\ a_{21} & a_{22} & a_{23} & -\cos \psi_2(y_f^{(O_2)} - \rho_w) \\ a_{31} & a_{32} & a_{33} & s_w \\ 0 & 0 & 0 & 1 \end{bmatrix} \mathbf{R}_w(h_w, \varepsilon_w)$$

(6.3.42)

wherein

$$a_{11} = -\cos \psi_2 \cos \gamma_{w2} \cos \phi_w + \sin \psi_2 \sin \phi_w \qquad (6.3.43)$$

$$a_{12} = \cos \psi_2 \cos \gamma_{w2} \sin \phi_w + \sin \psi_2 \cos \phi_w \qquad (6.3.44)$$

$$a_{13} = \cos \psi_2 \sin \gamma_{w2} \qquad (6.3.45)$$

$$a_{21} = -\sin \psi_2 \cos \gamma_{w2} \cos \phi_w - \cos \psi_2 \sin \phi_w \qquad (6.3.46)$$

$$a_{22} = \sin \psi_2 \cos \gamma_{w2} \sin \phi_w - \cos \psi_2 \cos \phi_w \qquad (6.3.47)$$

$$a_{23} = \sin \psi_2 \sin \gamma_{w2} \qquad (6.3.48)$$

$$a_{31} = \sin \gamma_{w2} \cos \phi_w \qquad (6.3.49)$$

$$a_{32} = -\sin \gamma_{w2} \sin \phi_w \qquad (6.3.50)$$

$$a_{33} = \cos \gamma_{w2} \qquad (6.3.51)$$

(ii) Position vector $\mathbf{r}_2(h_w, v_w, \phi_w, s_w)$ is represented in Cartesian coordinates as

$$\boldsymbol{\rho}_2(h_w, v_w, \phi_w, s_w) = \begin{bmatrix} a_{11} & a_{12} & a_{13} \\ a_{21} & a_{22} & a_{23} \\ a_{31} & a_{32} & a_{33} \end{bmatrix} \boldsymbol{\rho}_w(h_w, v_w) + \begin{bmatrix} \sin \psi_2(y_f^{(O_2)} - \rho_w) \\ -\cos \psi_2(y_f^{(O_2)} - \rho_w) \\ s_w \end{bmatrix}$$

$$= \mathbf{L}_{2w}(\phi_w, s_w)\boldsymbol{\rho}_w(h_w, v_w) + \mathbf{R} \qquad (6.3.52)$$

Here, $\mathbf{L}_{2w}$ is a 3×3 matrix obtained from $\mathbf{M}_{2w}$. Matrix $\mathbf{L}_{2w}$ may be obtained as

$$\mathbf{L}_{2w} = \mathbf{L}_{2f}\mathbf{L}_{fs}\mathbf{L}_{sw} \tag{6.3.53}$$

whereas vector $\mathbf{R}$ is defined as

$$\mathbf{R} = \begin{bmatrix} \sin \psi_2(y_f^{(O_2)} - \rho_w) & -\cos \psi_2(y_f^{(O_2)} - \rho_w) & s_w \end{bmatrix}^T \tag{6.3.54}$$

(iii) Considering s_w as constant ($s_w = c$), the relative velocity of the worm-thread surface with respect to gear-tooth surface may be obtained as

$$\mathbf{v}_{2,s_w=c}^{(w2)} = \dot{\boldsymbol{\rho}}_2 = \dot{\mathbf{L}}_{2w}\boldsymbol{\rho}_w + \dot{\mathbf{R}} \tag{6.3.55}$$

wherein

$$\dot{\mathbf{L}}_{2w} = \dot{\mathbf{L}}_{2f}\mathbf{L}_{fs}\mathbf{L}_{sw} + \mathbf{L}_{2f}\mathbf{L}_{fs}\dot{\mathbf{L}}_{sw} \tag{6.3.56}$$

and

$$\dot{\phi}_w = -\frac{\cos \beta_c}{\rho_w \cos \lambda_w}\dot{x}_f^{(O_c)} \tag{6.3.57}$$

Then, the equation of meshing $f_1^{(w2)}$ may be obtained as

$$f_1^{(w2)} = \mathbf{n}_2 \cdot \mathbf{v}_{2,s_w=c}^{(w2)} = 0 \tag{6.3.58}$$

wherein

$$\mathbf{n}_2 = \mathbf{L}_{2w}\mathbf{n}_w \tag{6.3.59}$$

Here, $\mathbf{n}_w$ is the unit normal to the worm-thread surface.

(iv) Considering ϕ_w as constant ($\phi_w = c$), the relative velocity of the worm-thread surface with respect to gear-tooth surface may be obtained as

$$\mathbf{v}_{2,\phi_w=c}^{(w2)} = \dot{\boldsymbol{\rho}}_2 = \dot{\mathbf{L}}_{2w}\boldsymbol{\rho}_w + \dot{\mathbf{R}} \tag{6.3.60}$$

wherein

$$\dot{\mathbf{L}}_{2w} = \dot{\mathbf{L}}_{2f}\mathbf{L}_{fs}\mathbf{L}_{sw} \tag{6.3.61}$$

and

$$\dot{s}_w = -\frac{1}{\tan \beta_c}\dot{x}_f^{(O_c)} \tag{6.3.62}$$

Then, the equation of meshing $f_2^{(w2)}$ may be obtained as

$$f_2^{(w2)} = \mathbf{n}_2 \cdot \mathbf{v}_{2,\phi_w=c}^{(w2)} = 0 \tag{6.3.63}$$

(v) Derivations of derivatives $\dot{\psi}_2$, $\dot{y}_f^{(O_2)}$, and $\dot{x}_f^{(O_2)}$ are

$$\dot{\psi}_2 = \left(\frac{d\theta_2}{d\theta_1} + \frac{d\mu_2}{d\theta_1} \right) \frac{d\theta_1}{dt} \tag{6.3.64}$$

$$\dot{y}_f^{(O_2)} = \frac{dy_f^{(O_2)}}{d\theta_1} \frac{d\theta_1}{dt} \tag{6.3.65}$$

$$\dot{x}_f^{(O_c)} = \frac{dx_f^{(O_c)}}{d\theta_1} \frac{d\theta_1}{dt} \tag{6.3.66}$$

Derivatives $\dfrac{d\theta_2}{d\theta_1}$, $\dfrac{d\mu_2}{d\theta_1}$, and $\dfrac{dy_f^{(O_2)}}{d\theta_1}$ are given by Eqs. (6.3.17), (6.3.18), and (6.3.21), respectively. Derivative $\dfrac{dx_f^{(O_c)}}{d\theta_1}$ is given by

$$\frac{dx_f^{(O_c)}}{d\theta_1} = -\rho_1 \left(1 - \frac{1}{\sqrt{1 - \varepsilon_1^2 \sin^2 \theta_1}} \right)$$

$$+\rho_1 \left(\frac{\varepsilon_1^2 \sin \theta_1 \cos \theta_1}{\sqrt{1 - \varepsilon_1^2 \sin^2 \theta_1}} - \varepsilon_1 \sin \theta_1 \right) \cos \mu_2$$

$$-r(\theta_2) \sin \mu_2 \varepsilon_1^2 \sin^2 \theta_1 \tag{6.3.67}$$

6.4 Generation of the Eccentric Gear Providing Localized Contact

Localization of bearing contact is achieved by double-crowning of the eccentric circular gear. Application of a grinding disk or a grinding worm may be applied as described in detail in Litvin and Fuentes (Litvin & Fuentes, 2004). A parabolic coefficient a_{pc} is applied for profile crowning whereas a parabolic coefficient a_{pl} is applied for longitudinal crowning of the eccentric gear-tooth surface.

TOOTH CONTACT ANALYSIS. The eccentric gear-tooth surface Σ_1 and the noncircular gear-tooth surface Σ_2 have been previously obtained in their own rigidly connected reference systems S_1 and S_2, respectively. A fixed reference system S_f is considered for investigation of tooth contact along the cycle of meshing. The algorithm of tooth contact analysis must be applied for each single pair of teeth.

The path of contact and function of transmission errors have been investigated for an eccentric gear with design parameters shown in Table 6.4.1.

Figure 6.4.1 shows paths of contact on tooth number 1 and 11 for various values of shaft angle error. Figure 6.4.2 shows the transmission function along a whole

Table 6.4.1. *Design parameters of eccentric gear drive.*

Number of teeth of the eccentric pinion, N_1	21
Number of teeth of the noncircular gear, N_2	63
Module, m	2.0 mm
Pressure angle, α	20°
Helix angle, β	15°
Face width	25 mm
Parameter of eccentricity, ε_1	0.2
Parabolic coefficient for profile crowning, a_{pc}	0.002 mm^{-1}
Radius of grinding disk, ρ_D	125.0 mm
Parabolic coefficient for longitudinal crowning, a_{lc}	0.0001 mm^{-1}

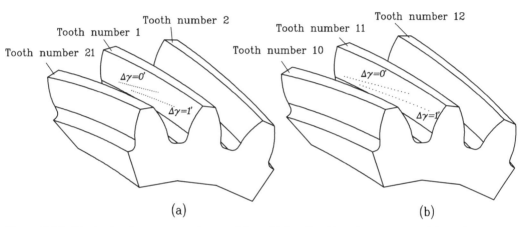

Figure 6.4.1. Contact paths on eccentric gear drive with localized bearing contact at (a) tooth number 1, and (b) tooth number 11, for several values of shaft-crossing angle error.

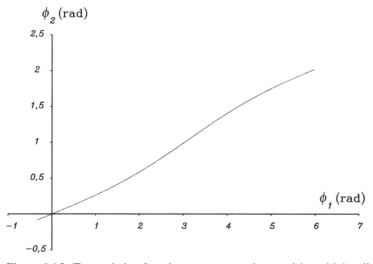

Figure 6.4.2. Transmission function at an eccentric gear drive with localized bearing contact.

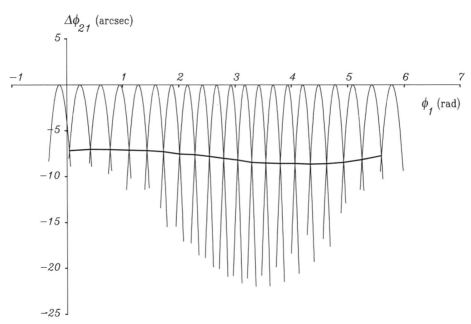

Figure 6.4.3. Function of transmission errors of an eccentric gear drive with localized bearing contact.

revolution of the eccentric gear obtained by tooth contact analysis. The theoretical transmission function may be derived as

$$\phi_{2t} - \phi_{20} = \int_{\phi_{10}}^{\phi_1} \left(\frac{c_1}{c_1 - \sqrt{1 - \varepsilon^2 \sin^2 \phi_1} + \varepsilon_1 \cos \phi_1} - 1 \right) d\phi_1 \qquad (6.4.1)$$

Figure 6.4.3 shows the function of transmission errors along a whole revolution of the eccentric gear, obtained as

$$\Delta\phi_{21} = (\phi_2 - \phi_{20}) - (\phi_{2t} - \phi_{20}) \qquad (6.4.2)$$

7 Design of Internal Noncircular Gears

7.1 Introduction

In this chapter, we consider the design of a gear drive formed by an external gear 1 and an internal noncircular gear 2. Henceforth, we assume that gear 1 is the driving one. Such a gear drive is called *internal noncircular gear drive* for the purpose of abbreviation.

The mentioned gear drive may be applied for variation of output speed of a gear mechanism and for function generation. The rotation of gears 1 and 2 of the drive are performed in the same direction, as shown in Fig. 7.1.1.

We may consider two cases of transformation of rotation by: (i) a pair of conjugated centrodes that form a pair of conjugated frictional disks, an *external* one (that is the smaller disk 1) and *internal* one (that is the larger disk 2); and (ii) a pair of gears, external 1 and internal 2. The frictional disks (and the gear centrodes 1 and 2) are in internal tangency.

7.2 Derivation of Centrodes

7.2.1 Preliminary Considerations of Kinematics of Internal Gear Drive

Figure 7.1.1(a) shows schematically that external gear 1 (centrode 1) and internal gear 2 (centrode 2) perform rotation about centers O_1 and O_2 in the same direction. Rotation about center O_1 of external gear 1 is represented by vector $\omega^{(1)}(\phi_1)$. Similarly, rotation about center O_2 of internal gear 2 is represented by vector $\omega^{(2)}(\phi_1)$. Here, ϕ_1 represents the angle of rotation of external gear 1 about center O_1.

The derivative function $m_{12}(\phi_1)$ is represented as

$$m_{12}(\phi_1) = \frac{\omega^{(1)}(\phi_1)}{\omega^{(2)}(\phi_1)} > 1 \tag{7.2.1}$$

The relative angular velocity is

$$\omega^{(12)}(\phi_1) = \omega^{(1)}(\phi_1) - \omega^{(2)}(\phi_1) = \omega^{(1)}(\phi_1)\left(1 - \frac{1}{m_{12}}\right) \tag{7.2.2}$$

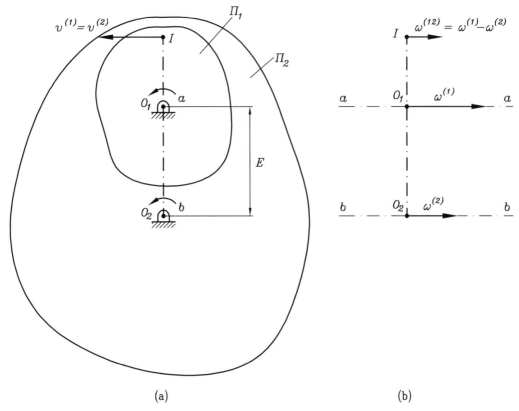

(a)   (b)

Figure 7.1.1. Illustration of transformation of rotation in case of internal noncircular gears:
(a) O_1 and O_2 are the centers of rotation; (b) vectors $\omega^{(1)}$ and $\omega^{(2)}$ are directed along the
respective axes of rotation.

wherein $\left(1 - \dfrac{1}{m_{12}}\right) < 1.$

Application of internal gear drives allows the relative angular velocity $\omega^{(12)}(\phi_1)$
and the relative velocity

$$\mathbf{v}^{(12)}(\phi_1) = \mathbf{v}^{(1)}(\phi_1) - \mathbf{v}^{(2)}(\phi_1) \tag{7.2.3}$$

to be reduced. Reduction of relative velocity means reduction of sliding. This is the
reason why internal gear drives are applied as gear drives with greater efficiency.

7.2.2 Basic Equations of Centrodes

The derivation of centrodes is based on the concept similar to the one applied in
Chapter 2. The centrodes of the internal gear drive are in tangency at point I that
belongs to the extended line $O_1 - O_2$ (Fig. 7.2.1), wherein

$$\mathbf{v}^{(12)} = \mathbf{0}, \quad \mathbf{v}^{(1)} = \mathbf{v}^{(2)} \tag{7.2.4}$$

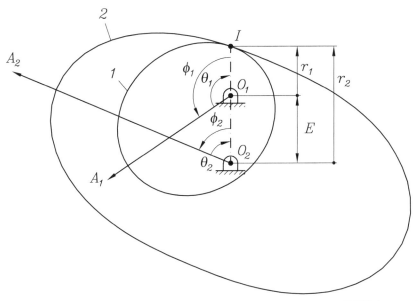

Figure 7.2.1. Illustration of mating centrodes 1 and 2, polar axes $\overline{O_i A_i}$ $(i = 1, 2)$, rotation angles ϕ_i $(i = 1, 2)$, polar angles θ_i $(i = 1, 2)$, and polar vectors r_i $(i = 1, 2)$.

Equation $\mathbf{v}^{(12)} = \mathbf{0}$ means that the centrodes of the internal gear drive roll over each other and the sliding velocity of I is zero. Point I is the instantaneous center of rotation of the centrodes in relative motion.

Considering as given the derivative function $m_{12}(\phi_1)$, and taking into account that $\mathbf{v}^{(12)} = \mathbf{0}$, we obtain

$$m_{12}(\phi_1) = \frac{\omega^{(1)}(\phi_1)}{\omega^{(2)}(\phi_1)} = \frac{d\phi_1}{d\phi_2} = \frac{r_2(\phi_1)}{r_1(\phi_1)} = \frac{E + r_1(\phi_1)}{r_1(\phi_1)} \tag{7.2.5}$$

Equation (7.2.5) allows the conjugated centrodes σ_1 and σ_2 to be derived and represented as functions of angle ϕ_1, where ϕ_1 is the angle of rotation of the driving centrode σ_1.

We may represent σ_1 and σ_2 as polar curves in terms of polar centrode angle θ_1, taking into account that $\theta_1 \equiv \phi_1$, but θ_1 is measured opposite to ϕ_1:

(i) Centrode σ_1 is represented as

$$r_1(\phi_1) = E \frac{1}{m_{12}(\phi_1) - 1} \tag{7.2.6}$$

or as

$$r_1(\theta_1) = E \frac{1}{m_{12}(\theta_1) - 1} \tag{7.2.7}$$

(ii) Centrode σ_2 is represented as

$$r_2(\phi_1) = E\frac{m_{12}(\phi_1)}{m_{12}(\phi_1) - 1} \tag{7.2.8}$$

$$\phi_2(\phi_1) = \int_0^{\phi_1} \frac{d\phi_1}{m_{12}(\phi_1)} \tag{7.2.9}$$

7.2.3 Design of Centrodes σ_1 and σ_2 as Closed-Form Curves

The conditions of centrodes σ_1 and σ_2 of an internal noncircular gear drive to be closed-form curves are similar to those discussed in Section 2.6 for an external gear drive.

We require for an internal gear drive observation of the following conditions:

(i) The derivative function $m_{12}(\phi_1)$ must be a periodic function with period $T = T_1/n_1$, where T_1 is the period of revolution of driving centrode σ_1 (gear 1) and n_1 is an integer number.
(ii) The ratio between the revolutions n_1 and n_2 of centrodes 1 and 2, respectively, must be

$$\frac{T_1}{n_1} = \frac{T_2}{n_2} \tag{7.2.10}$$

where n_1 and n_2 are integer numbers.
(iii) The gear center distance E must be determined as a function of n_1 and n_2.

Derivations similar to those performed in Section 2.6 yield the following equation of E for an internal noncircular gear drive:

$$\frac{2\pi}{n_2} = \int_0^{\frac{2\pi}{n_1}} \frac{r_1(\phi_1)}{E + r_1(\phi_1)} d\phi_1 \tag{7.2.11}$$

where $r_1(\phi_1)$ is the known polar equation of driving centrode σ_1.

Generally, the solution of Eq. (7.2.11) for E may be obtained numerically and is an iterative process. An analytical solution of (7.2.11) is obtained for a gear drive with elliptical and modified elliptical centrodes (see following).

7.3 Examples of Design of Internal Noncircular Gear Drives

7.3.1 Gear Drive with Elliptical Pinion

Step 1. Centrode σ_1 of the drive is represented in polar form (see Section 4.3.2) as

$$r_1(\phi_1) = \frac{a(1 - e^2)}{1 - e\cos\phi_1}, \quad 0 \le \phi_1 \le 2\pi \tag{7.3.1}$$

Step 2. The derivative function of the drive, $m_{21}(\phi_1) = \dfrac{1}{m_{12}(\phi_1)}$, is represented (see Eq. (7.2.5)) as

$$m_{21}(\phi_1) = \frac{r_1(\phi_1)}{E + r_1(\phi_1)} = \frac{p}{E + p - Ee\cos\phi_1} \tag{7.3.2}$$

where $p = a(1 - e^2)$.

Step 3. The next step is determination of center distance $E = E(a, e, n)$. It is based on application of

$$\frac{2\pi}{n} = \int_0^{2\pi} m_{21}(\phi_1)d\phi_1 = \int_0^{2\pi} \frac{p}{E + p - Ee\cos\phi_1}d\phi_1 \tag{7.3.3}$$

The meaning of this equation is the requirement that the pinion (centrode σ_1) will perform n revolutions for one revolution of gear 2.

Simple transformation of Eq. (7.3.3) yields

$$\frac{\pi}{n} = \frac{p}{E + p} \int_0^{\pi} \frac{d\phi_1}{1 - q\cos\phi_1} \tag{7.3.4}$$

where $q = \dfrac{Ee}{E + p}$.

It follows from the tables of integrals (see Dwight, 1961, Problem 858.524), that

$$\int_0^{\pi} \frac{d\phi_1}{1 - q\cos\phi_1} = \frac{\pi}{\sqrt{1 - q^2}} \tag{7.3.5}$$

Equations (7.3.4) and (7.3.5) yield the following solution for E:

$$E(a, e, n) = a[-1 + \sqrt{1 + (n^2 - 1)(1 - e^2)}] \tag{7.3.6}$$

It is remarkable, as shown later, that Eq. (7.3.6) works as well for determination of the center distance of a gear drive with modified elliptical or oval pinion.

Step 4. Centrode σ_2 of gear 2 is represented as

$$r_2(\phi_1) = E + r_1(\phi_1) \tag{7.3.7}$$

$$\phi_2(\phi_1) = \int_0^{\phi_1} m_{21}d\phi_1 = \frac{p}{E + p} \int_0^{\phi_1} \frac{d\phi_1}{1 - q\cos\phi_1} \tag{7.3.8}$$

Transmission function $\phi_2(\phi_1)$ may be determined numerically by integration of Eq. (7.3.8), or analytically (see Dwight, 1961, Problem 446.00) using

$$\int \frac{d\phi_1}{1 - q\cos\phi_1} = \frac{2}{\sqrt{1 - q^2}} \arctan \frac{1 + q\tan\frac{\phi_1}{2}}{\sqrt{1 - q^2}} \tag{7.3.9}$$

Figures 7.3.1(a) and (b) show the gear drives with elliptical pinion determined for $n = 2$ and $n = 3$, respectively; $a = 39.3378$ mm, $e = 0.5$.

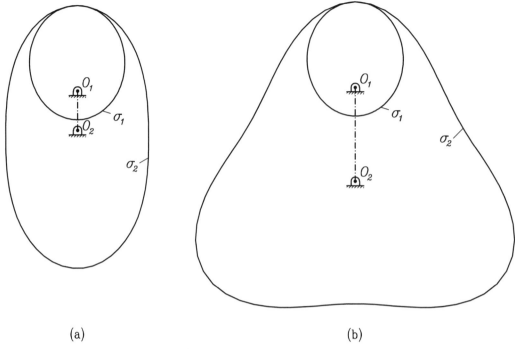

(a)   (b)

Figure 7.3.1. Illustration of elliptical centrode with parameters $a = 39.3378$ mm and $e = 0.5$, and the internal noncircular centrode in case of (a) $n = 2$ and (b) $n = 3$.

7.3.2 Gear Drive with Modified Elliptical Pinion

Step 1. The modified elliptical centrode σ_1 of gear 1 (see Section 4.3.4) consists of two branches, the upper and the lower, and is represented by

$$r_1^{(I)}(\phi_1) = \frac{a(1 - e^2)}{1 - e\cos(m_I\phi_1)}, \quad 0 \le \phi_1 \le \pi/m_I \tag{7.3.10}$$

$$r_1^{(II)}(\phi_1) = \frac{a(1 - e^2)}{1 - e\cos[m_{II}(2\pi - \phi_1)]}, \quad \pi/m_I \le \phi_1 \le 2\pi \tag{7.3.11}$$

where $m_{II} = \frac{m_I}{2m_I - 1}$ (see Eq. (4.3.35)).

Step 2. The derivative function of the drive, $m_{21}(\phi_1)$, is determined as

$$m_{21}(\phi_1) = \frac{\omega^{(2)}}{\omega^{(1)}} = \frac{r_1(\phi_1)}{r_2(\phi_1)} = \frac{r_1(\phi_1)}{E + r_1(\phi_1)} \tag{7.3.12}$$

Equations (7.3.10), (7.3.11), and (7.3.12) yield

$$m_{21}^{(I)}(\phi_1) = \frac{p}{p + E - Ee\cos(m_I\phi_1)}, \quad 0 \le \phi_1 \le \frac{\pi}{m_I} \tag{7.3.13}$$

$$m_{21}^{(II)}(\phi_1) = \frac{p}{p + E - Ee\cos(m_{II}(2\pi - \phi_1))}, \quad \frac{\pi}{m_I} \le \phi_1 \le 2\pi \tag{7.3.14}$$

Step 3. Gear 1 (centrode σ_1) performs n revolutions for one revolution of gear 2, and therefore

$$\frac{2\pi}{n} = \int_0^{\frac{\pi}{m_I}} m_{21}^{(I)} d\phi_1 + \int_{\frac{\pi}{m_I}}^{2\pi} m_{21}^{(II)} d\phi_1 \tag{7.3.15}$$

We recall that centrodes σ_1 and σ_2 are closed curves and n is an integer number.

Step 4. We apply Eqs. (7.3.13), (7.3.14), and (7.3.15) for determination of $E = E(a, e, n)$. The derivation of E is based on the following procedure:

(i) The variables of the integrals of Eq. (7.3.15) are changed as

$$m_I \phi_1 = x, \quad m_{II}(2\pi - \phi_1) = y \tag{7.3.16}$$

(ii) We take into account that (see Dwight, 1961, Problem 858.524)

$$\int_0^\pi \frac{dx}{1 - q\cos x} = \frac{\pi}{\sqrt{1 - q^2}}, \quad \int_0^\pi \frac{dy}{1 - \mu\cos y} = \frac{\pi}{\sqrt{1 - \mu^2}} \tag{7.3.17}$$

We then obtain the solution $E = E(a, e, n)$ for the drive with modified elliptical gears that is the same as for the drive with conventional ellipses represented by Eq. (7.3.6).

Step 5. Centrode σ_2 is represented by

$$r_2^{(I)}(\phi_1) = E + r_1^{(I)}(\phi_1) \tag{7.3.18}$$

$$\phi_2^{(I)}(\phi_1) = \int_0^{\phi_1} m_{21}^{(I)}(\phi_1) d\theta_1, \quad 0 \le \phi_1 \le \frac{\pi}{m_I} \tag{7.3.19}$$

Similarly, we represent equations for $r_2^{(II)}(\phi_1)$ and $\phi_2^{(II)}(\phi_1)$.

Determination of $\phi_2^{(i)}(\phi_1)$, $(i = I, II)$, may be obtained by using the expression for independent integral $\int \frac{dx}{1 - q\cos x}$ (see Dwight, 1961, Problem 446.00).

Figure 7.3.2 illustrates designed internal gear drives with modified elliptical gears.

7.3.3 Gear Drive with Oval Pinion

Step 1. The centrode σ_1 of the pinion is an oval (see Fig. 7.3.3), a curve obtained as the result of transformation of a conventional ellipse performed as follows: (i) the magnitude of the radius vector of the ellipse is observed, but (ii) the position angle is increased twice.

The polar equation of the oval gear 1 is represented as

$$r_1(\phi_1) = \frac{a(1 - e^2)}{1 - e\cos 2\phi_1}, \quad 0 \le \phi_1 \le 2\pi \tag{7.3.20}$$

wherein a and e are the semi-length of the major axis and the eccentricity of the ellipse being transformed.

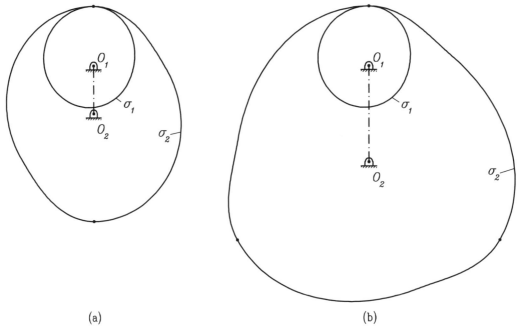

(a)   (b)

Figure 7.3.2. Illustration of modified elliptical centrode with parameters $a = 37.1241$ mm and $e = 0.2$, $m_I = 1.5$, and $m_{II} = 0.75$, and the internal noncircular centrode in case of (a) $n = 2$ and (b) $n = 3$.

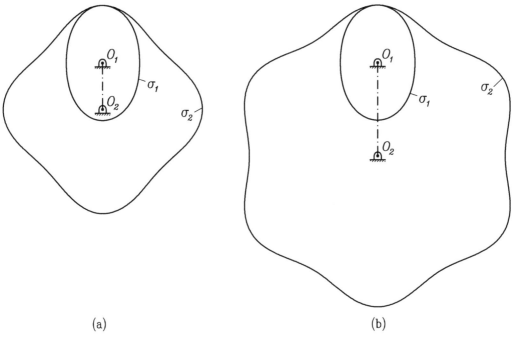

(a)   (b)

Figure 7.3.3. Illustration of oval centrode (with parameters $a = 37.1241$ mm and $e = 0.2$), and the internal noncircular centrode in case of (a) $n = 2$ and (b) $n = 3$.

Step 2. The radius vector of the internal gear 2 is $r_2(\phi_1) = E + r_1(\phi_1)$ (wherein E is the shortest center distance) and the derivative function $m_{21}(\phi_1)$ is represented as

$$m_{21}(\phi_1) = \frac{r_1(\phi_1)}{E + r_1(\phi_1)} = \frac{p}{E + p - Ee\cos 2\phi_1} \qquad (7.3.21)$$

Step 3. Relation between rotations of gears 1 and 2. This relation is similar to the one that has been represented for an external gear drive (see Section 4.3.5.3):

$$2\pi = n\int_0^{2\pi} \frac{p}{E + p - Ee\cos 2\phi_1} d\phi_1 = \left(\frac{np}{E+p}\right)\int_0^{2\pi} \frac{d\phi_1}{1 - \left(\frac{Ee}{E+p}\right)\cos 2\phi_1}$$
$$(7.3.22)$$

Step 4. Derivation of center distance E. The variable ϕ_1 is changed for x wherein

$$2\phi_1 = x, \quad d\phi_1 = \frac{dx}{2} \qquad (7.3.23)$$

that yields

$$2\pi = \left(\frac{np}{E+p}\right)\left(\frac{4}{2}\right)\int_0^{\pi} \frac{dx}{1 - q\cos x}, \quad q = \frac{Ee}{E+p} \qquad (7.3.24)$$

From tables of integrals (Dwight, 1961),

$$\int_0^{\pi} \frac{dx}{1 - q\cos x} = \frac{\pi}{\sqrt{1 - q^2}} \qquad (7.3.25)$$

After simple derivations, we obtain equation $E = E(a, e, n)$ that is the same as Eq. (7.3.6) represented for internal noncircular gear drives with elliptical pinion and modified elliptical pinion.

Step 5. Centrode σ_2 of internal gear is determined with

$$r_2(\phi_1) = E + r_1(\phi_1) \qquad (7.3.26)$$

$$\phi_2 = \int_0^{\phi_1} m_{21}(\phi_1)d\phi_1 \qquad (7.3.27)$$

where $m_{21}(\phi_1)$ is represented by Eq. (7.3.21).

Transmission function $\phi_2(\phi_1)$ may be determined by numerical integration of Eq. (7.3.27), and analytically using the work of Dwight (Dwight, 1961, see Problem 446.00), with $a^2 > b^2$.

Centrodes σ_1 and σ_2 of the drive are illustrated by Fig. 7.3.3.

7.3.4 Gear Drive with Eccentric Pinion

Step 1. Derivation of $r_1(\phi_1) = |\overline{O_1 M}|$. Centrode σ_1 of gear 1 is an eccentric circle of radius a and eccentricity $e = |\overline{O_1 C_1}| = |\overline{O_1 C_2}|$ (Fig. 7.3.4). Centrode σ_1 (the eccentric circle) performs rotation about center O_1 and is shown in two positions: the initial one, and the current one after rotation on angle ϕ_1.

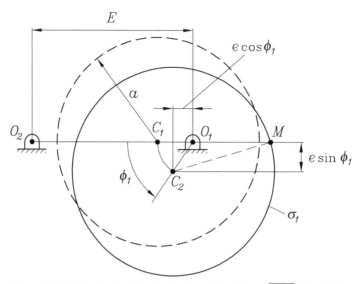

Figure 7.3.4. Illustration for derivation of $r_1(\phi_1) = |\overline{O_1 M}|$; O_1 is the center of rotation of eccentric circle of radius a; C_1 and C_2 are initial and current positions of geometric center of circle a.

The conjugated centrode σ_2 (not shown in Fig. 7.3.4) is an internal curve and performs rotation about center O_2; $E = |\overline{O_2 O_1}|$ is the shortest center distance.

Centrodes σ_1 and σ_2 are in tangency at a point M that belongs to the extended line $O_1 - O_2$. Current point M of tangency of centrodes moves along $O_1 - O_2$ in the process of rotation. It follows from the drawings that

$$r_1(\phi_1) = |\overline{O_1 M}| = (a^2 - e^2 \sin^2 \phi_1)^{0.5} - e \cos \phi_1, \quad 0 \le \phi_1 \le 2\pi \quad (7.3.28)$$

Step 2. Derivative function $m_{21}(\phi_1)$. The derivative function $m_{21}(\phi_1)$ is determined as

$$m_{21}(\phi_1) = \frac{d\phi_2}{d\phi_1} = \frac{r_1(\phi_1)}{r_2(\phi_2)} = \frac{r_1(\phi_1)}{E + r_1(\phi_1)}$$

$$= 1 - \frac{c}{c + (1 - \varepsilon^2 \sin^2 \phi_1)^{0.5} - \varepsilon \cos \phi_1} \quad (7.3.29)$$

where $c = \dfrac{E}{a}$ and $\varepsilon = \dfrac{e}{a}$.

Step 3. Derivation of $c = E/a$. Centrode σ_2 (of internal gear) must be a closed-form curve, and this may be obtained by the observation of

$$\frac{2\pi}{n} = \int_0^{2\pi} \left(1 - \frac{c}{c + (1 - \varepsilon^2 \sin^2 \phi_1)^{0.5} - \varepsilon \cos \phi_1}\right) d\phi_1 \quad (7.3.30)$$

The meaning of Eq. (7.3.30) is that centrode σ_1 (of the pinion) performs n revolutions for one revolution of centrode σ_2 of internal gear; n is an integer number.

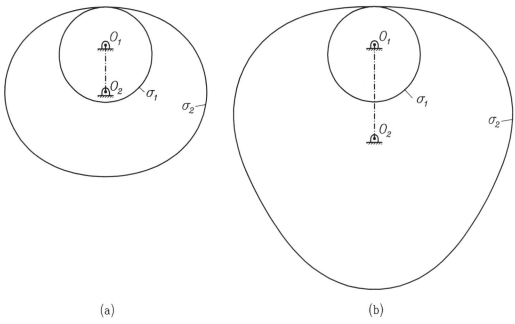

(a)   (b)

Figure 7.3.5. Illustration of an eccentric centrode with parameters $a = 36.75$ mm and $e = 7.35$ mm, and the internal noncircular centrode in case of (a) $n = 2$ and (b) $n = 3$.

Solution of Eq. (7.3.30) for c require iterations and numerical integration. Simple transformations yield

$$\frac{n-1}{n} = \frac{1}{2\pi} \int_0^\pi \frac{c}{c + (1 - \varepsilon^2 \sin^2 \phi_1)^{0.5} - \varepsilon \cos \phi_1} d\phi_1$$

$$+ \frac{1}{2\pi} \int_\pi^{2\pi} \frac{c}{c + (1 - \varepsilon^2 \sin^2 \phi_1)^{0.5} - \varepsilon \cos \phi_1} d\phi_1$$

$$= \frac{1}{\pi} \int_0^\pi \frac{c}{c + (1 - \varepsilon^2 \sin^2 \phi_1)^{0.5} - \varepsilon \cos \phi_1} d\phi_1 \qquad (7.3.31)$$

Step 4. Derivation of centrode σ_2. The centrode σ_2 is represented by

$$r_2(\phi_1) = E + r_1(\phi_1) \qquad (7.3.32)$$

$$\phi_2(\phi_1) = \int_0^{\phi_1} m_{21}(\phi_1) d\phi_1$$

$$= \phi_1 - \int_0^{\phi_1} \frac{E}{E + (a^2 - e^2 \sin^2 \phi_1)^{0.5} - e \cos \phi_1} d\phi_1, \quad 0 \le \phi_1 \le 2\pi n$$

$$(7.3.33)$$

Figure 7.3.5 illustrates examples of derived internal drives.

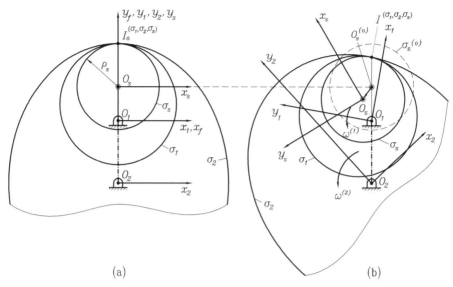

Figure 7.4.1. Illustration of movable coordinate systems S_1, S_2, S_s, and fixed coordinate system S_f: (a) initial position of centrodes; (b) current positions of centrodes.

7.4 Generation of Planar Internal Noncircular Gears by Shaper

Generation of pinion 1 with centrode σ_1 of the internal gear drive may be performed by a rack cutter, a shaper, or a hob (see Chapter 5).

The generation of internal noncircular gear 2 with centrode σ_2 must be performed by a shaper s. The generated profile Γ_2 of gear 2 is determined as the envelope to the family of profiles Γ_s of the shaper in relative motion of Γ_s with respect to Γ_2. The conjugation of pinion 1 and gear 2 is provided by considering simultaneous tangency of centrodes of 1, 2, and shaper s (Fig. 7.4.1).

The generation of Γ_2 by Γ_s may be accompanied by undercutting that is a defect of meshing of Γ_2 and Γ_s. Such a defect is the result of *interference* of Γ_s in the space of Γ_2. Design of shaper s with a larger pressure angle α_s is in favor of avoidance of undercutting, but proper geometric relations between Γ_s, Γ_2, and Γ_1 are preferable.

Henceforth, we will consider the following cases of tangency of rolling centrodes: (i) tangency of centrodes σ_1, σ_2, and σ_s (Fig. 7.4.1), and (ii) tangency of centrodes σ_2 and σ_s (Fig. 7.4.2).

TANGENCY OF CENTRODES σ_1, σ_2, AND σ_s. Figure 7.4.1(a) shows that the centrodes are initially in tangency at common point $I_o^{(\sigma_1,\sigma_2,\sigma_s)}$. Three movable coordinate systems S_1, S_2, and S_s are considered that are rigidly connected to pinion 1, internal gear 2, and shaper s. Coordinate systems S_1 and S_2 perform related rotations about O_1 and O_2, respectively (Fig. 7.4.1); points $I_o^{(\sigma_1,\sigma_2,\sigma_s)}$ and $I^{(\sigma_1,\sigma_2,\sigma_s)}$ are the initial and current positions of centrodes tangency (Figs. 7.4.1(a) and (b)) and they are the instantaneous centers of rotation of centrodes in relative motion.

Point $I^{(\sigma_1,\sigma_2,\sigma_s)}$ of tangency of centrodes σ_1 and σ_2 moves in the process of motion along line $O_2 - O_1 - I^{(\sigma_1,\sigma_2,\sigma_s)}$ (Fig. 7.4.1). Tangency of σ_s with σ_1 and σ_2 at

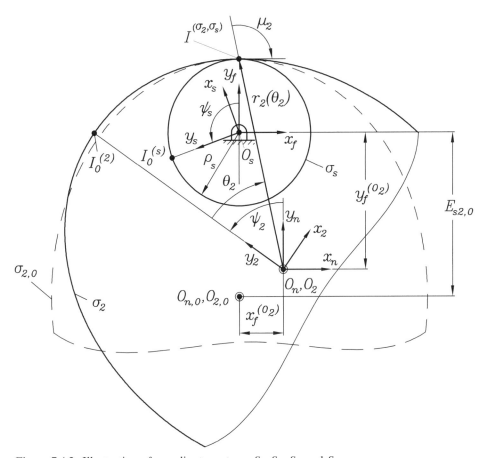

Figure 7.4.2. Illustration of coordinate systems S_n, S_f, S_s, and S_2.

point $I^{(\sigma_1,\sigma_2,\sigma_s)}$ is obtained wherein shaper s performs a complex motion: (a) translational motion with coordinate system S_s, and (b) rotational motion about O_s; $O_s^{(o)}$ and O_s are the initial and current positions of the origin of S_s (Fig. 7.4.1(b)). Coordinate system S_f is the fixed one.

Centrodes σ_1, σ_2, and σ_s roll over each other in the process of transformation of motion. The common normal to profiles Γ_1, Γ_2, and Γ_s (not shown in Fig. 7.4.1) passes through the point of tangency $I^{(\sigma_1,\sigma_2,\sigma_s)}$ of the three centrodes, and this is the condition of conjugation of profiles Γ_1, Γ_2, and Γ_s.

TANGENCY OF CENTRODES σ_2 AND σ_s. Shaper s with profile Γ_s generates profile Γ_2 of internal gear 2 wherein coordinate systems S_s and S_2 are rigidly connected to s and 2. We apply the following coordinate systems (Fig. 7.4.2):

(i) Coordinate system S_s, that is rigidly connected to s and performs rotation about O_s on angle ψ_s;

(ii) Coordinate system S_2, that is rigidly connected to gear 2 and performs (a) translational motion with auxiliary coordinate system S_n (Fig. 7.4.2), and (b) rotational motion about origin O_n of S_n on angle ψ_2 (Fig. 7.4.2).

The translational motions of S_n and S_2 are performed collinear to axes (x_f, y_f). Figure 7.4.2 shows the current point $I^{(\sigma_2,\sigma_s)}$ of tangency of centrodes σ_2 and σ_s. Initially, centrodes σ_2 and σ_s were in tangency at point $I_o^{(\sigma_1,\sigma_2,\sigma_s)}$ (Fig. 7.4.1(a)). Then, σ_2 and σ_s become in tangency at point $I^{(\sigma_2,\sigma_s)}$ (Fig. 7.4.2), and their common tangent forms angle μ_2 with radius vector $\overline{O_2 I}$ of centrode 2 and is perpendicular to $\overline{O_s I}$.

Tangency of σ_2 and σ_s at $I^{(\sigma_2,\sigma_s)}$ is provided, if centrode σ_2 with coordinate system S_n is translated on magnitudes $x_f^{(O_2)}$ and $E_{s2,0} + y_f^{(O_2)}$ along axes (x_f, y_f) of S_f. Notice that the magnitude of $y_f^{(O_2)}$ expressed in S_f will be negative. Here, $E_{s2,0}$ is the initial center distance between centrodes σ_2 and σ_s (see Fig. 7.4.2); magnitudes $x_f^{(O_2)}$ and $y_f^{(O_2)}$ may be represented as functions of polar angle θ_2 as

$$x_f^{(O_2)} = -r_2(\theta_2)\cos\mu_2 \tag{7.4.1}$$

$$y_f^{(O_2)} = +r_2(\theta_2)\sin\mu_2 + \rho_s \tag{7.4.2}$$

$$\psi_2 = \theta_2 + \mu_2 - \frac{\pi}{2} \tag{7.4.3}$$

Centrodes σ_s and σ_2 will have a common normal at $I^{(\sigma_2,\sigma_s)}$ that forms angle μ_2 with position vector $\overline{O_2 I}$ (see the notation of angle μ in Section 2.5). The common tangent to σ_s and σ_2 will pass through point $I^{(\sigma_2,\sigma_s)}$ of tangency of σ_2 and σ_s, and the relative velocity $\mathbf{v}^{(s2)}$ will satisfy

$$\mathbf{v}^{(s2)} \cdot \mathbf{N}^{(s)} = \mathbf{v}^{(s2)} \cdot \mathbf{N}^{(2)} = 0 \tag{7.4.4}$$

where $\mathbf{N}^{(s)}$ and $\mathbf{N}^{(2)}$ are the normals to centrodes σ_s and σ_2.

Observation of Eq. (7.4.4) is the condition of derivation of the equation of meshing of shaper σ_s and internal gear 2.

PROFILE Γ_s OF SHAPER s. Figure 7.4.3 shows the generating involute profile Γ_s of the shaper teeth in coordinate system S_s. Involute profile of the shaper may be obtained as well considering the generating process of a spur gear by a rack cutter (Litvin & Fuentes, 2004). In matrix form, the involute profile is represented as

$$\mathbf{r}_s(\theta_s) = \begin{bmatrix} \rho_{bs}\left[\sin(\theta_s - \theta_{0s}) - \theta_s\cos(\theta_s - \theta_{0s})\right] \\ \rho_{bs}\left[\cos(\theta_s - \theta_{0s}) + \theta_s\sin(\theta_s - \theta_{0s})\right] \\ 0 \\ 1 \end{bmatrix} \tag{7.4.5}$$

Here, $\rho_{bs} = \rho_s\cos\alpha_s$ is the base radius of the shaper, ρ_s is its pitch radius, and α_s is the pressure angle. Parameter θ_{0s} for a standard involute shaper is determined as

$$\theta_{0s} = \frac{\pi m}{4\rho_s} + \mathrm{inv}\alpha_s \tag{7.4.6}$$

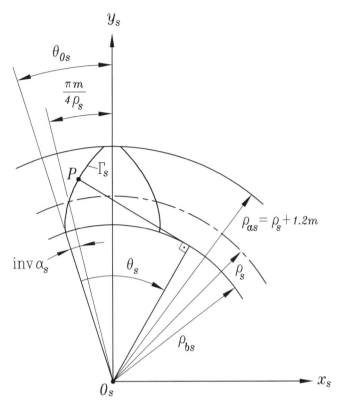

Figure 7.4.3. Representation of shaper profile Γ_s in coordinate system S_s.

where m is the module of the gear drive and $\mathrm{inv}\,\alpha_s = \tan\alpha_s - \alpha_s$. The unit normal $\mathbf{n}_s(\theta_s)$ to the profile Γ_s is represented by

$$\mathbf{n}_s(\theta_s) = \frac{\dfrac{d\mathbf{r}_s}{d\theta_s} \times \mathbf{k}}{\left|\dfrac{d\mathbf{r}_s}{d\theta_s}\right|} \qquad (7.4.7)$$

MATRIX DERIVATION OF PROFILE Γ_2 OF INTERNAL GEAR 2. Profile Γ_2 is generated as the envelope to the family of generating profiles Γ_s determined in S_2 by

$$\mathbf{r}_2(\theta_s, \theta_2) = \mathbf{M}_{2n}(\psi_2)\mathbf{M}_{nf}(x_f^{(O_2)}, y_f^{(O_2)})\mathbf{M}_{fs}(\psi_s)\mathbf{r}_s(\theta_s) \qquad (7.4.8)$$

Matrices $\mathbf{M}_{2n}$, $\mathbf{M}_{nf}$, and $\mathbf{M}_{fs}$ of Eq. (7.4.8) describe coordinate transformation from coordinate system S_s to coordinate system S_2. Here,

$$\mathbf{M}_{2n}(\psi_2) = \begin{bmatrix} \cos\psi_2 & \sin\psi_2 & 0 & 0 \\ -\sin\psi_2 & \cos\psi_2 & 0 & 0 \\ 0 & 0 & 1 & 0 \\ 0 & 0 & 0 & 1 \end{bmatrix} \qquad (7.4.9)$$

$$\mathbf{M}_{nf}(x_f^{(O_2)}, y_f^{(O_2)}) = \begin{bmatrix} 1 & 0 & 0 & -x_f^{(O_2)} \\ 0 & 1 & 0 & -y_f^{(O_2)} \\ 0 & 0 & 1 & 0 \\ 0 & 0 & 0 & 1 \end{bmatrix} \tag{7.4.10}$$

$$\mathbf{M}_{fs}(\psi_s) = \begin{bmatrix} \cos\psi_s & -\sin\psi_s & 0 & 0 \\ \sin\psi_s & \cos\psi_s & 0 & 0 \\ 0 & 0 & 1 & 0 \\ 0 & 0 & 0 & 1 \end{bmatrix} \tag{7.4.11}$$

Matrix transformation Eq. (7.4.8) may be expressed by matrices (3×3) as

$$\boldsymbol{\rho}_2(\theta_s, \theta_2) = \mathbf{L}_{2n}\mathbf{L}_{nf}\mathbf{L}_{fs}\boldsymbol{\rho}_s(\theta_s) + \mathbf{R}$$

$$= \begin{bmatrix} \cos\psi_2 & \sin\psi_2 & 0 \\ -\sin\psi_2 & \cos\psi_2 & 0 \\ 0 & 0 & 1 \end{bmatrix} \begin{bmatrix} 1 & 0 & 0 \\ 0 & 1 & 0 \\ 0 & 0 & 1 \end{bmatrix} \begin{bmatrix} \cos\psi_s & -\sin\psi_s & 0 \\ \sin\psi_s & \cos\psi_s & 0 \\ 0 & 0 & 1 \end{bmatrix} \boldsymbol{\rho}_s(\theta_s) + \mathbf{R} \tag{7.4.12}$$

where $\mathbf{R}$ is given by

$$\mathbf{R} = \begin{bmatrix} -x_f^{(O_2)}\cos\psi_2 - y_f^{(O_2)}\sin\psi_2 \\ x_f^{(O_2)}\sin\psi_2 - y_f^{(O_2)}\cos\psi_2 \\ 0 \end{bmatrix} \tag{7.4.13}$$

Profile Γ_2 is determined by simultaneous consideration of Eq. (7.4.12) and the scalar product (see Litvin & Fuentes, 2004):

$$\mathbf{n}_2 \cdot \mathbf{v}_2^{(s2)} = \mathbf{L}_{2s}\mathbf{n}_s \cdot \dot{\boldsymbol{\rho}}_2 = f(\theta_s, \theta_2) = 0 \tag{7.4.14}$$

Here, $\mathbf{n}_s$ is the unit normal to the shaper and $\mathbf{v}_2^{(s2)}$ is the relative velocity, which is represented in S_2 by

$$\mathbf{v}_2^{(s2)} = \dot{\boldsymbol{\rho}}_2 = (\dot{\mathbf{L}}_{2n}\mathbf{L}_{nf}\mathbf{L}_{fs} + \mathbf{L}_{2n}\mathbf{L}_{nf}\dot{\mathbf{L}}_{fs})\boldsymbol{\rho}_s + \dot{\mathbf{R}} \tag{7.4.15}$$

wherein

$$\dot{\mathbf{L}}_{2n} = \begin{bmatrix} -\sin\psi_2 & \cos\psi_2 & 0 \\ -\cos\psi_2 & -\sin\psi_2 & 0 \\ 0 & 0 & 1 \end{bmatrix} \dot{\psi}_2 \tag{7.4.16}$$

$$\dot{\mathbf{L}}_{fs} = \begin{bmatrix} -\sin\psi_s & -\cos\psi_s & 0 \\ \cos\psi_s & -\sin\psi_s & 0 \\ 0 & 0 & 1 \end{bmatrix} \dot{\psi}_s \tag{7.4.17}$$

$$\dot{\mathbf{R}} = \begin{bmatrix} -\dot{x}_f^{(O_2)} \cos \psi_2 - \dot{y}_f^{(O_2)} \sin \psi_2 + (x_f^{(O_2)} \sin \psi_2 - y_f^{(O_2)} \cos \psi_2)\dot{\psi}_2 \\ \dot{x}_f^{(O_2)} \sin \psi_2 - \dot{y}_f^{(O_2)} \cos \psi_2 + (x_f^{(O_2)} \cos \psi_2 + y_f^{(O_2)} \sin \psi_2)\dot{\psi}_2 \\ 0 \end{bmatrix} \quad (7.4.18)$$

Derivatives $\dot{\psi}_2, \dot{\psi}_s, \dot{x}_f^{(O_2)}, \dot{y}_f^{(O_2)}$ are obtained as

$$\dot{\psi}_2 = \frac{d\psi_2}{d\theta_2}\dot{\theta}_2 = \left(1 + \frac{d\mu_2}{d\theta_2}\right)\dot{\theta}_2 \quad (7.4.19)$$

$$\dot{\psi}_s = \frac{d\psi_s}{d\theta_2}\dot{\theta}_2 = \frac{1}{\rho_s}\frac{ds(\theta_2)}{d\theta_2}\dot{\theta}_2 \quad (7.4.20)$$

$$\dot{x}_f^{(O_2)} = \left(-\frac{dr_2(\theta_2)}{d\theta_2}\cos \mu_2 + r_2(\theta_2)\sin \mu_2 \frac{d\mu_2}{d\theta_2}\right)\dot{\theta}_2 \quad (7.4.21)$$

$$\dot{y}_f^{(O_2)} = \left(-\frac{dr_2(\theta_2)}{d\theta_2}\sin \mu_2 - r_2(\theta_2)\cos \mu_2 \frac{d\mu_2}{d\theta_2}\right)\dot{\theta}_2 \quad (7.4.22)$$

Here, $\frac{dr_2}{d\theta_2}$ is obtained as

$$\frac{dr_2}{d\theta_2} = \frac{dr_2}{d\theta_1}\frac{d\theta_1}{d\theta_2} = \frac{dr_1}{d\theta_1}\frac{E + r_1}{r_1} \quad (7.4.23)$$

where

$$r_2(\theta_1) = E + r_1(\theta_1) \quad (7.4.24)$$

$$\theta_2(\theta_1) = \int_0^{\theta_1} \frac{r_1(\theta_1)}{E + r_1(\theta_1)}d\theta_1 \quad (7.4.25)$$

Term $\frac{d^2r_2}{d\theta_2^2}$ (which is needed for $\frac{d\mu_2}{d\theta_2}$) is obtained by differentiation of Eq. (7.4.23) as

$$\frac{d^2r_2}{d\theta_2^2} = \frac{d^2r_1}{d\theta_1^2}\frac{(E + r_1)^2}{r_1^2} - \left(\frac{dr_1}{d\theta_1}\right)^2 \frac{E(E + r_1)}{r_1^3} \quad (7.4.26)$$

Term μ_2 is obtained as (see Eq. (2.5.1))

$$\mu_2 = \arctan \frac{r_2(\theta_2)}{\dfrac{dr_2}{d\theta_2}} \quad (7.4.27)$$

Term $\frac{d\mu_2}{d\theta_2}$ is obtained as (see Eq. (2.8.4))

$$\frac{d\mu_2}{d\theta_2} = \cos^2 \mu_2 \left[1 - \frac{r_2(\theta_2)\dfrac{d^2r_2}{d\theta_2^2}}{\left(\dfrac{dr_2}{d\theta_2}\right)^2}\right] \quad (7.4.28)$$

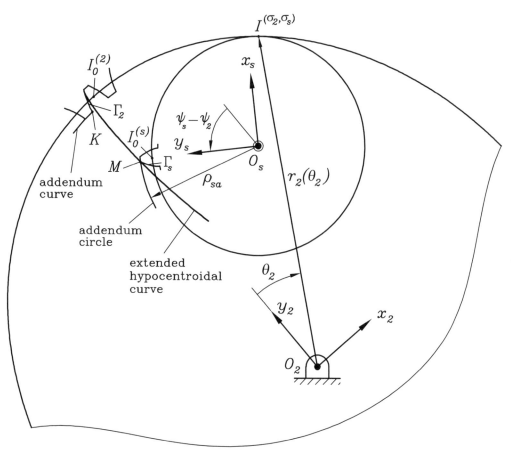

Figure 7.5.1. Illustration of the trajectory described by the tip of the shaper tooth during the relative motion between the shaper and the internal noncircular gear.

Term $\dfrac{ds(\theta_2)}{d\theta_2}$ is obtained as (see Eq. (2.8.1))

$$\frac{ds(\theta_2)}{d\theta_2} = \frac{r_2(\theta_2)}{\sin \mu_2} \qquad (7.4.29)$$

In all previous derivations, expressions $r_1(\theta_1)$, $\dfrac{dr_1(\theta_1)}{d\theta_1}$, and $\dfrac{d^2 r_1(\theta_1)}{d\theta_1^2}$ are known and depend on the type of centrode (elliptical, eccentric, modified elliptical).

7.5 Conditions of Nonundercutting of Planar Internal Noncircular Gears

Undercutting of profile Γ_2 of planar internal noncircular gear tooth is caused by the interference with the profile of the tip of the profile Γ_s of the generating shaper during the relative motion between the shaper and the internal noncircular gear (see Fig. 7.5.1). The tip of profile Γ_s is represented by point M, which belongs to the addendum circle of the shaper.

The limiting condition of undercutting corresponds to positions wherein the trajectory of point M reaches point K, which belongs to the addendum curve of the internal noncircular gear tooth profile. The trajectory of point M depends on the method of generation of the internal noncircular gear by the shaper. Here, *axial* generation is considered, which means that center distance between the shaper and the internal noncircular gear is modified according to expressions of $x_f^{(O_2)}$ and $y_f^{(O_2)}$ (see Eqs. (7.4.1) and (7.4.2)).

Two approaches are considered for determination of the limiting value of the pitch radius ρ_s^* of the shaper for avoidance of undercutting. In consequence, the limiting value of shaper teeth, N_s^*, is obtained as

$$N_s^* = \frac{2\rho_s^*}{m} \tag{7.5.1}$$

where m is the module of the gear drive.

7.5.1 Approach A

The limiting condition of undercutting can be represented numerically by a set of nonlinear equations. Solution of such a set of nonlinear equations implies determination of the limiting value of the pitch radius ρ_s^*. The proposed approach is based on the following algorithm:

(1) Point M belongs to the addendum circle of the shaper, and this condition is represented by

$$f_1(\theta_s^M, \rho_s^*) = |\mathbf{r}_s^{(M)}(\theta_s^M, \rho_s^*)| - (\rho_s^* + 1.2m) = 0 \tag{7.5.2}$$

Here, $|\mathbf{r}_s^{(M)}(\theta_s^M, \rho_s^*)| = \sqrt{(r_{sx}^M)^2 + (r_{sy}^M)^2}$, and vector $\mathbf{r}_s^{(M)}(\theta_s^M, \rho_s^*)$ is represented by Eq. (7.4.5) wherein $\rho_s = \rho_s^*$ is considered as a variable.

(2) Point K belongs to the profile Γ_2 of the noncircular gear tooth, and this condition implies that equation of meshing Eq. (7.4.14) is satisfied at point K,

$$f_2(\theta_s^K, \theta_2^K, \rho_s^*) = 0 \tag{7.5.3}$$

(3) Point K belongs to the addendum curve of the noncircular gear, and this condition is represented by

$$f_3(\theta_s^K, \theta_2^K, \rho_s^*, \theta_{\sigma_2}^K) = |\mathbf{r}_2^{(K)}(\theta_s^K, \theta_2^K, \rho_s^*)| - (r_2(\theta_{\sigma_2}^K) - m) = 0 \tag{7.5.4}$$

Here,

(i) $|\mathbf{r}_2^{(K)}(\theta_s^K, \theta_2^K, \rho_s^*)| = \sqrt{(r_{2x}^K)^2 + (r_{2y}^K)^2}$ where vector $\mathbf{r}_2^{(K)}(\theta_s^K, \theta_2^K, \rho_s^*)$ is obtained by matrix transformation as

$$\mathbf{r}_2^{(K)}(\theta_s^K, \theta_2^K, \rho_s^*) = \mathbf{M}_{2s}(\theta_2^K)\mathbf{r}_s(\theta_s^K, \rho_s^*) \tag{7.5.5}$$

(ii) $r_2(\theta_{\sigma_2}^K)$ is the amplitude of polar vector of the centrode σ_2 corresponding to point K and $\theta_{\sigma_2}^K$ is its polar angle.

(4) Polar angle $\theta_{\sigma_2}^K$ satisfies

$$f_4(\theta_s^K, \theta_2^K, \theta_{\sigma_2}^K, \rho_s^*) = \tan\theta_{\sigma_2}^K - \frac{r_{2x}^K(\theta_s^K, \theta_2^K, \rho_s^*)}{r_{2y}^K(\theta_s^K, \theta_2^K, \rho_s^*)} = 0 \qquad (7.5.6)$$

(5) The trajectory of point M in coordinate system S_2 may be obtained by application of matrix transformation

$$\mathbf{r}_2^{(M)}(\theta_s^M, \rho_s^*, \theta_2) = \mathbf{M}_{2s}(\theta_2)\mathbf{r}_s(\theta_s^M, \rho_s^*) \qquad (7.5.7)$$

wherein θ_2 is the parameter of the curve that represents the trajectory of M.

(6) Intersection of curve represented by Eq. (7.5.7) and profile Γ_2 at point K represented by Eq. (7.5.5) provides a vectorial equation

$$\mathbf{r}_2^{(M)}(\theta_s^M, \rho_s^*, \theta_2) - \mathbf{r}_2^{(K)}(\theta_s^K, \theta_2^K, \rho_s^*) = \mathbf{0} \qquad (7.5.8)$$

that yields two additional scalar equations

$$f_5(\theta_s^M, \theta_s^K, \theta_2^K, \rho_s^*, \theta_2) = r_{2x}^M - r_{2x}^K = 0 \qquad (7.5.9)$$

$$f_6(\theta_s^M, \theta_s^K, \theta_2^K, \rho_s^*, \theta_2) = r_{2y}^M - r_{2y}^K = 0 \qquad (7.5.10)$$

(7) A system of six nonlinear equations f_1, f_2, f_3, f_4, f_5, f_6 is obtained as

$$f_1(\theta_s^M, \rho_s^*) = 0 \qquad (7.5.11)$$

$$f_2(\theta_s^K, \theta_2^K, \rho_s^*) = 0 \qquad (7.5.12)$$

$$f_3(\theta_s^K, \theta_2^K, \rho_s^*, \theta_{\sigma_2}^K) = 0 \qquad (7.5.13)$$

$$f_4(\theta_s^K, \theta_2^K, \theta_{\sigma_2}^K, \rho_s^*) = 0 \qquad (7.5.14)$$

$$f_5(\theta_s^M, \theta_s^K, \theta_2^K, \rho_s^*, \theta_2) = 0 \qquad (7.5.15)$$

$$f_6(\theta_s^M, \theta_s^K, \theta_2^K, \rho_s^*, \theta_2) = 0 \qquad (7.5.16)$$

and may be solved numerically for determination of unknowns $(\theta_s^M, \rho_s^*, \theta_s^K, \theta_2^K, \theta_{\sigma_2}^K, \theta_2)$.

7.5.2 Approach B

Approach B is based on the results provided in Table 7.5.1, developed by Litvin *et al.* (Litvin *et al.*, 1994), to determine the maximal number of teeth for various pressure angles in the case of internal circular involute gears (see Chapter 11 of Litvin & Fuentes, 2004).

Figure 7.5.2 shows the substitution of centrode σ_2 of the internal noncircular gear by a new centrode σ_{2c} of a circular gear with pitch radius equal to the minimal radius of the centrode σ_2. Centrode σ_{2c} is rigidly connected to system S_{2c}. The trajectory of point M in system S_{2c} is an extended hypocycloid.

Table 7.5.1. *Maximal number of shaper teeth for internal circular involute gears.*

Pressure angle	Generation method	Gear teeth	Shaper teeth
$\alpha_c = 20°$	Axial	$25 \leq N_2 \leq 31$	$N_c \leq 0.82 N_2 - 3.20$
	Axial	$32 \leq N_2 \leq 200$	$N_c \leq 1.004 N_2 - 9.162$
	Two-parameter	$36 \leq N_2 \leq 200$	$N_c \leq N_2 - 17.6$
$\alpha_c = 25°$	Axial	$17 \leq N_2 \leq 31$	$N_c \leq 0.97 N_2 - 5.40$
	Axial	$32 \leq N_2 \leq 200$	$N_c \leq N_2 - 6.00$
	Two-parameter	$23 \leq N_2 \leq 200$	$N_c \leq N_2 - 11.86$
$\alpha_c = 30°$	Axial	$15 \leq N_2 \leq 200$	$N_c \leq N_2 - 4.42$
	Two-parameter	$17 \leq N_2 \leq 200$	$N_c \leq N_2 - 8.78$

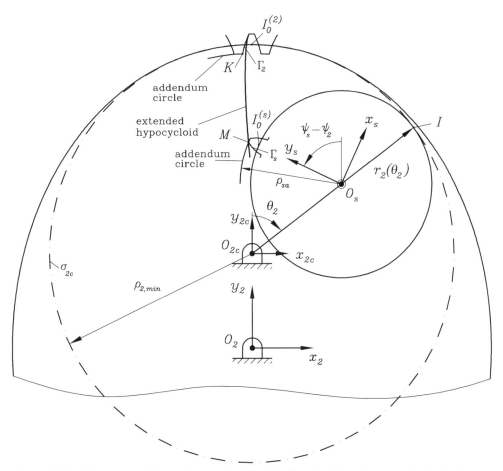

Figure 7.5.2. Illustration of the trajectory described by the tip of the shaper tooth during the relative motion between the shaper and the equivalent internal circular gear.

Table 7.5.2. *Gear data.*

Number of teeth of the elliptical pinion, N_1	31
Number of teeth of the internal noncircular gear, N_2	62
Number of revolutions of the elliptical pinion, n	2
Module, m	4.0 mm
Pressure angle, α	20.0°
Eccentricity of elliptical centrode σ_1, e	0.2

Approach B is based on the following algorithm:

(1) The minimal curvature radius of centrode σ_2 of the internal noncircular gear, $\rho_{2,min}$, is obtained by consideration of (see Eq. (2.8.8))

$$\rho_2(\theta_2) = \frac{\left[r_2(\theta_2)^2 + \left(\dfrac{dr_2}{d\theta_2} \right)^2 \right]^{3/2}}{r_2(\theta_2)^2 + 2 \left(\dfrac{dr_2}{d\theta_2} \right)^2 - r_2(\theta_2) \dfrac{d^2 r_2}{d\theta_2^2}} \tag{7.5.17}$$

$$\frac{d\rho_2}{d\theta_2} = 0 \tag{7.5.18}$$

Expressions $r_2(\theta_2)$, $\dfrac{dr_2(\theta_2)}{d\theta_2}$, and $\dfrac{d^2 r_2(\theta_2)}{d\theta_2^2}$ have been represented in Section 7.4.

(2) The number of teeth for the equivalent internal circular gear is obtained as

$$N_2^* = \frac{2\rho_{2,min}}{m} \tag{7.5.19}$$

(3) The limiting value of the number of shaper teeth can be obtained by application of Table 7.5.1 (see Chapter 11 of Litvin & Fuentes, 2004).

7.5.3 Numerical Example

An internal noncircular gear drive based on a conventional elliptical pinion is considered. The gear data are shown in Table 7.5.2.

Parameter a of the elliptical centrode σ_1 is obtained as (see Eq. (4.3.26))

$$a = \frac{N_1 m \pi}{4 \int_0^{\pi/2} \sqrt{1 - e^2 \sin^2 \theta_1} d\theta_1} = 62.631089 \text{ mm}$$

Table 7.5.3. *Solution for Approach A.*

Profile parameter of point M, θ_s^M	0.517701 rad
Limiting value of the pitch radius of the shaper, ρ_s^*	82.541688 mm
Profile parameter of point K, θ_s^K	0.172244 rad
Generalized parameter of motion for point K, θ_2^K	−0.139680 rad
Polar angle for polar vector of point K, $\theta_{\sigma_2}^K$	−0.031313 rad
Parameter of the trajectory of point M, θ_2	0.620299 rad

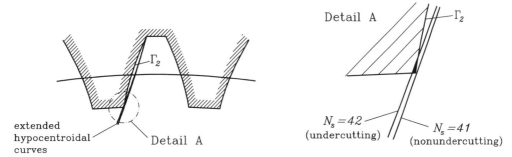

Figure 7.5.3. Observation of undercutting when $N_s = 42$ and nonundercutting when $N_s = 41$.

Center distance E of the gear drive is obtained as (see Eq. (7.3.6))

$$E = a[-1 + \sqrt{1 + (n^2 - 1)(1 - e^2)}] = 60.737850 \text{ mm}$$

APPLICATION OF APPROACH A. The solution of the set of the nonlinear equations corresponding to the Approach A provides the values of magnitudes (θ_s^M, ρ_s^*, θ_s^K, θ_2^K, $\theta_{\sigma_2}^K$, θ_2) shown in Table 7.5.3.
The limiting value of the number of shaper teeth is obtained as

$$N_s^* = \frac{2\rho_s^*}{m} = 41.2708$$

Figure 7.5.3 shows the trajectories of point M in two cases: for $N_s = 41$ ($\rho_s = 82$ mm) and for $N_s = 42$ ($\rho_s = 84$ mm). The detail A of Fig. 7.5.3 shows that undercutting occurs for $N_s = 42$, whereas undercutting does not occur for $N_s = 41$.

APPLICATION OF APPROACH B. The minimal radius of curvature of centrode σ_2 is obtained at $\theta_2 = 0$ (see Fig. 7.3.1(a)). In this case,

$$\rho_{2,min} = \frac{1}{\dfrac{1}{E + a(1 + e)} + \dfrac{e}{a(1 - e^2)}} = 93.589402 \text{ mm}$$

The number of teeth of the equivalent internal circular gear, N_2^*, is obtained as

$$N_2^* = \frac{2\rho_{2,min}}{m} = 46.794701$$

Applications of results shown in Table 7.5.1 (see Chapter 11 of Litvin & Fuentes, 2004) provides the following limiting value of the number of shaper teeth:

$$N_s^* = 1.004 N_2^* - 9.162 = 37.8199$$

Approach B provides a more conservative result than the exact solution provided by Approach A.

Application for Design of Planetary Gear
Train with Noncircular and Circular Gears

8.1 Introduction

A planetary gear train with noncircular and circular gears (Litvin & Ketov, 1949)
may be applied in design for the following purposes:

(1) For providing a lesser pressure angle for each pair of applied gears of the discussed train.
(2) For design of a function $y(x)$, $x_2 \geq x \geq x_1$, for which derivative y' varies its sign
in the mentioned interval. The design function will be obtained as the output of
the planetary gear train. We recall that function $y(x)$ with varied sign of $y'(x)$
cannot be generated by a pair of noncircular gears (see Section 10.3).
(3) For obtaining as the output of the planetary train a product $f_1(x) \cdot f_2(x)$ of two
nonlinear functions.

8.2 Structure and Basic Kinematic Concept of Planetary Train

We limit the discussions to a planetary gear train formed by two pairs of gears being
in external meshing. The first set of gears is formed by gears 1 and 2, and the second
set by gears 3 and 4 (Fig. 8.2.1). One of the gears, say gear 1, is held at rest. Gears 2
and 3 are called *satellites*. They are mounted on the carrier c and have two degrees
of freedom (i) in the motion with the carrier, and (ii) in relative motion, rotation
about the carrier.

Notation $\phi_{41}^{(c)}$ indicates that transformation of rotation is performed from gear 1
to gear 4 wherein carrier c is held at rest, and gears 1 and 2, 3 and 4 form a conventional gear train (not a planetary one). We then have

$$\phi_{41}^{(c)} = \phi_4(\phi_3(\phi_2(\phi_1))) \tag{8.2.1}$$

Here, $\phi_3 \equiv \phi_2$ because gears 3 and 2 do not perform rotation with respect to each
other. Another notation, $\phi_{4c}^{(1)}$, indicates that transformation of rotation is performed
from carrier c to gear 4 wherein gear 1 is held at rest. It follows from the kinematics
of a planetary gear train that

$$\phi_{4c}^{(1)} = \phi_{41}^{(c)} - \phi_1 \tag{8.2.2}$$

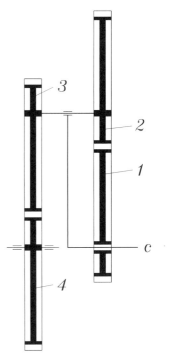

Figure 8.2.1. Structure of planetary train with two pairs of gears and carrier c.

Equation (8.2.2) may be obtained as follows:

(1) Consider that all four gears and the carrier perform rotation being rotated as a rigid body. The angle of rotation is ϕ_c and the positions of the gears and the carrier are determined by ϕ_c.

(2) However, because gear 1 is fixed, it is necessary to return gear 1 to the initial position. This is obtained by rotation of gear 1 in the direction that is opposite to the rotation performed in stage 1. The angle of rotation is $|\phi_1| = |\phi_c|$. At stage 2, the rotations of gears 1, 2, 3, and 4 are performed about their axes and the positions of the gears are determined by Eq. (8.2.2).

The final positions of the gears and the carrier obtained by rotation at steps 1 and 2 are determined by Eq. (8.2.2). This yields Eq. (8.2.2), which may be represented as

$$\phi_{4c}^{(1)} = \phi_{41}^{(c)} - \phi_1 = \phi_4(\phi_3(\phi_2(\phi_1))) - \phi_1 \tag{8.2.3}$$

Equation (8.2.3) has been derived considering that gears 1 and 2, 3 and 4, are in external meshing. The discussed approach may be extended for the case wherein gears of the planetary train have a mixed meshing, external and internal.

8.3 Planetary Gear Train with Elliptical Gears

Figure 8.3.1(a) shows a planetary gear with two pairs of identical elliptical gears 1 and 2, 3 and 4. The purpose of application of two pairs of gears is increasing the variation of the ratio of angular velocities of the input and output links of the train,

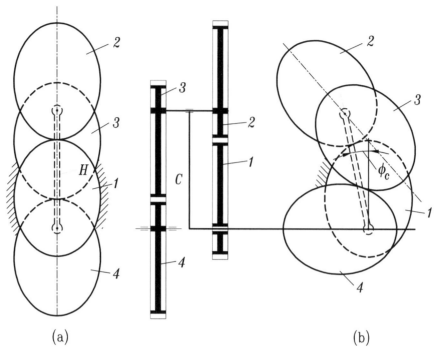

(a)   (b)

Figure 8.3.1. Planetary gear train with two pairs of elliptical gears: (a) initial position of carrier c and elliptical gears 1, 2, and 3, 4; (b) position of gears 2, 3, 4 and carrier c after rotation of carrier on angle ϕ_c whereas gear 1 is held at rest.

but with observation of limitation of pressure angle α_{12} for each pair of two elliptical gears (see Section 3.4).

The derivation of transmission function $\phi_{41}^{(c)}$ (see Eq. (8.2.1)) is performed as follows:

(i) Transmission function $\phi_2(\phi_1)$ is determined as (see Section 4.3.2.1)

$$\tan\left(\frac{\phi_2}{2}\right) = \left(\frac{1+e}{1-e}\right)\tan\frac{\phi_1}{2} \qquad (8.3.1)$$

(ii) Similarly, we obtain that

$$\tan\left(\frac{\phi_4}{2}\right) = \left(\frac{1+e}{1-e}\right)\tan\frac{\phi_3}{2} \qquad (8.3.2)$$

We take into account in further derivations that $\phi_3 \equiv \phi_2$. Gears 2 and 3 are mounted on the carrier and perform rotation with respect to the carrier with the same angular velocity.

(iii) Equations (8.3.1) and (8.3.2) yield

$$\phi_{41}^{(c)} = 2\arctan\left[\left(\frac{1+e}{1-e}\right)^2\tan\left(\frac{\phi_1}{2}\right)\right] \qquad (8.3.3)$$

The meaning of Eq. (8.3.3) is that application of two pairs of elliptical gears is equivalent to application of one pair of elliptical gears with a very large eccentricity.

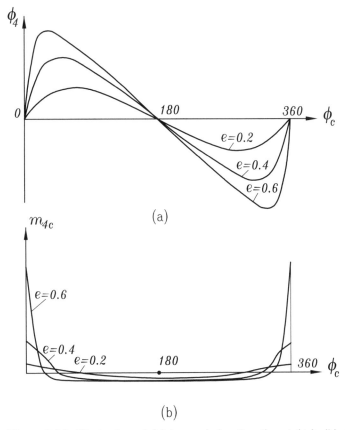

Figure 8.3.2. Illustration of (a) transmission function $\phi_4(\phi_c)$; (b) derivative function $m_{4c} = \omega_4/\omega_c$.

However, large eccentricity is accompanied with a large pressure angle (see Section 3.4), but this is avoided by application of a planetary gear train with two pairs of elliptical gears.

Drawings of Fig. 8.3.2(a) show that the output of the planetary train (function $\phi_4(\phi_c)$) is a function of reverse motion of gear 4 performed while the carrier is rotated in the same direction. Figure 8.3.2(b) shows that the derivative function $m_{4c}(\phi_c)$ varies its sign in the interval of generation. Such variation (of the sign) is impossible for generation of a function by application of one pair of noncircular gears (see Chapter 10).

8.4 Planetary Gear Train with Noncircular and Circular Gears

It will be represented in Chapter 10 that generation of a function $y(x)$ requires that the derivative $y'(x)$ must be positive in the interval of generation. This obstacle may be avoided if instead of $y(x)$ the following function is generated:

$$y_1(x) = y(x) + bx \tag{8.4.1}$$

with observation $y_1'(x) > 0$ in the interval of generation.

To obtain as the output function $y(x)$ with a varied sign of derivative $y'(x)$, it becomes necessary the subtraction of linear function

$$y_2(x) = bx \qquad (8.4.2)$$

from function $y_1(x)$. Such subtraction has been performed by application of a spiral bevel gear differential for derivation of $y(x) = \sin x$ (see Chapter 10).

Similar generation may be accomplished by application of a planetary gear train as follows:

(1) Assume that noncircular gears 1 and 2 (Fig. 8.2.1) generate the function Eq. (8.4.1). Circular gears 3 and 4 are designed with gear ratio 1, and $\phi_{41}^{(c)}$ is proportional to $y_1(x) = y(x) + bx$.

(2) Taking into account Eq. (8.2.2), we obtain that the output of the planetary train, $\phi_{4c}^{(1)}$, is proportional to function $y(x) = \sin x$.

9 Transformation of Rotation into Translation with Variation of Gear Ratio

9.1 Introduction

Meshing of a noncircular gear with a rack is considered in this book in two different ways: (i) wherein the rack is a generating tool, and (ii) the rack and the noncircular gear form a mechanism for transformation of motion.

Figure 9.1.1 illustrates that for previous case (ii), rotation with angular velocity ω_1 is transformed into translation with velocity v_2 and the ratio ω_1/v_2 varies. The mechanism applied for such transformation of motion (Fig. 9.1.1) is formed by a noncircular gear 1 and rack 2. The centrodes 1 and 2 are curves determined by function $s_2(\phi_1)$, denoted as $F(\phi_1)$.

The instantaneous point I of tangency of centrodes (Fig. 9.1.1) is determined as follows:

(1) The derivative $F'(\phi_1)$ of function $s_2(\phi_1) = F(\phi_1)$ is obtained as

$$F'(\phi_1) = \frac{ds_2}{d\phi_1} = \left(\frac{ds_2}{dt}\right)\left(\frac{dt}{d\phi_1}\right) = \frac{v_2}{\omega_1} \tag{9.1.1}$$

where $v_2 = \dfrac{ds_2}{dt}$ is the instantaneous velocity of translation of rack 2, and $\dfrac{d\phi_1}{dt}$ is the instantaneous velocity of rotation of noncircular gear 1.

(2) The instantaneous center I of rotation of gear 1 and rack 2 (in relative motion) is located at a point of line $\overline{O_1 n}$. Here, O_1 is the center of rotation of the gear 1, and $\overline{O_1 n}$ is drawn perpendicular to velocity $\mathbf{v}_2$ of translation. The relative velocity $\mathbf{v}_{12} = \mathbf{v}_1 - \mathbf{v}_2$ is equal to zero at point I.

(3) The location $|\overline{O_1 I}|$ of the instantaneous center of rotation on line $\overline{O_1 n}$ is determined as

$$\mathbf{v}_2 = \mathbf{v}_1 = \omega_1 \times \overline{O_1 I}$$

We then obtain that

$$|\overline{O_1 I}| = \frac{v_2}{\omega_1} = F'(\phi_1) \tag{9.1.2}$$

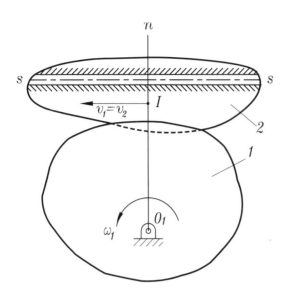

Figure 9.1.1. Centrodes 1 and 2 of a noncircular gear and a rack.

9.2 Determination of Centrodes of Noncircular Gear and Rack

Centrodes 1 and 2 are traced out by the instantaneous center I of rotation in coordinate systems S_1 and S_2 that are rigidly connected to gear 1 and rack 2, respectively.

(i) Centrode 1 is a polar curve determined as

$$r_1(\phi_1) = F'(\phi_1) \tag{9.2.1}$$

The polar axis is $\overline{O_1 A_1}$ (Fig. 9.2.1(a)), and $r_1(\phi_1)$ is the current position vector of centrode 1.

(ii) Centrode 2 (Fig. 9.2.1(b)) is represented in coordinate system (x_2, y_2) by

$$x_2 = \int_0^{\phi_1} r_1(\phi_1)d\phi, \qquad y_2 = r_1(\phi_1) - r_1^{(0)}(\phi_1^{(0)}) \tag{9.2.2}$$

9.3 Application of Mechanism Formed by a Noncircular Gear and Rack

Consider that the mechanism formed by a noncircular gear and a rack shown in Fig. 9.1.1 is applied for generation of function $y = f(x)$, $x_2 \geq x \geq x_1$ as follows:

(i) Angle ϕ_1 of rotation of noncircular gear and translation s_2 of rack are represented as

$$\phi_1 = \kappa_1(x - x_1) \tag{9.3.1}$$

$$s_2 = \kappa_2(y - y_1) = \kappa_2[f(x) - f(x_1)] \tag{9.3.2}$$

where κ_1 and κ_2 are scalar coefficients. Equations (9.3.1) and (9.3.2) determine function $s_2(\phi_1)$ parametrically.

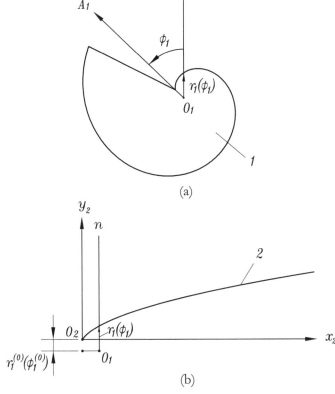

Figure 9.2.1. Toward derivation of centrodes 1 and 2: (a) centrode 1 of noncircular gear; (b) centrode 2 of rack.

(ii) Function of gear ratio is obtained as

$$F'(\phi_1) = \frac{ds_2}{d\phi_1} = \left(\frac{ds_2}{dx}\right)\left(\frac{dx}{d\phi_1}\right) = \frac{\kappa_2}{\kappa_1} f'(x) \tag{9.3.3}$$

(iii) Gear centrode (Fig. 9.2.1(a)) is determined parametrically as

$$\phi_1 = \kappa_1(x - x_1) \tag{9.3.4}$$

$$r_1(x) = \frac{\kappa_2}{\kappa_1} f'(x) \tag{9.3.5}$$

(iv) The centrode of the rack is determined in coordinate system (x_2, y_2) as

$$x_2(x) = \kappa_2[f(x) - f(x_1)] \tag{9.3.6}$$

$$y_2(x) = r_1 - r_{10} = \frac{\kappa_2}{\kappa_1}[f'(x) - f'(x_1)] \tag{9.3.7}$$

Axis $x_2(x)$ coincides with the direction of translational motion of the rack. We recall that derivative $f'(x)$ of the generated function $y(x)$ is to be a smooth and differentiable function.

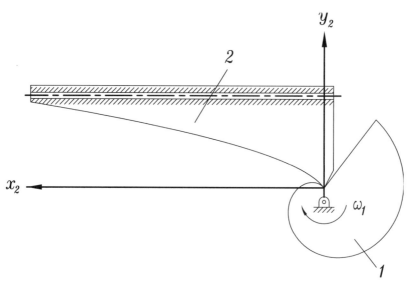

Figure 9.3.1. Illustration of centrodes 1 and 2 determined for generation of function $y(x) = \kappa_3 x^2$.

NUMERICAL EXAMPLE 9.3.1. Consider generation of function $y(x) = \kappa_3 x^2$, $x_a \geq x \geq x_b$, where $x_a = 200$ m, $x_b = 1{,}200$ m, $\kappa_3 = 2.5 \times 10^{-6}$ s/m. Centrode 1 is an unclosed curve. Coefficient $\kappa_1 = 0.4 \cdot (\pi/360)$ 1/m; $\kappa_2 = 1/12$ 1/s. The units of x and y are in meters (m) and seconds (s), respectively.

The angle of rotation of gear and displacement s_2 of the rack are represented as

$$\phi_1(x) = \kappa_1(x - x_a) \qquad (9.3.8)$$

$$s_2(y) = \kappa_2(y - y(x_a)) \qquad (9.3.9)$$

where κ_1 and κ_2 are the assigned coefficients. The gear centrode is represented by

$$\phi_1(x) = \kappa_1(x - x_a) \qquad (9.3.10)$$

$$r_1(x) = \frac{ds_2}{d\phi_1} = \frac{2\kappa_2\kappa_3}{\kappa_1}x \qquad (9.3.11)$$

It is easy to verify that $r_1(\phi_1)$ is an Archimedes spiral.

The centrode of the rack is represented in coordinate system (x_2, y_2) by

$$x_2 = \kappa_2\kappa_3(x^2 - x_a^2) \qquad (9.3.12)$$

$$y_2 = r_1(x) - r(x_a) = \frac{2\kappa_2\kappa_3}{\kappa_1}(x - x_a) \qquad (9.3.13)$$

Centrodes 1 and 2 are represented in Fig. 9.3.1.

10 Tandem Design of Mechanisms for Function Generation and Output Speed Variation

10.1 Introduction

In the past, generation of functions based on synthesis of planar linkages was the subject of research of many distinguished researchers based on synthesis of planar linkages (Artobolevski *et al.*, 1959; Burmester, 1888; Chebyshev, 1955; Freudenstein, 1955). The success of modern technology for the manufacture of noncircular gears has inspired researchers to apply noncircular gears for generation of functions (Dooner, 2001; Litvin *et al.*, 2008; Ottaviano *et al.*, 2008; Modler *et al.*, 2009; Litvin *et al.*, 2009).

The contents of this chapter cover generation of functions by a single pair of noncircular gears and a new approach for generation of functions by application of a multigear drive with noncircular gears. It follows from the kinematics of gear generation that function $y(x)$, $x_1 \leq x \leq x_2$, to be generated must be a monotonous increasing function with derivative $y'(x) > 0$. In the case wherein the derivative $y'(x)$ is of a varied sign in the interval of derivation, an algorithm for function generation is represented (see Section 10.3).

The application of a multigear drive with noncircular gears for function generation has the following advantages: (i) obtaining an integrated impact for the output results, (ii) favorable conditions of force transmission, and (iii) performance of synthesis with a number of design parameters substantially increased.

The specific characteristic of the proposed approach for application of a multigear drive is illustrated as follows:

(a) Figure 10.1.1 shows the centrodes of a gear drive wherein only a single pair of gears is applied (Fig. 10.1.1(a)) for generation of function $\psi(\alpha)$ (Fig. 10.1.1(b)).
(b) Figure 10.1.2(a) shows the case wherein two pairs of gears are applied for generation of function $\psi(\alpha)$ (10.1.2(b)). We then have that (i) the transmission function of centrodes 1 and 2 is $\beta(\alpha) \equiv g_1(\alpha)$, and (ii) transmission function of gears 3 and 4 is $\delta(\gamma) \equiv g_2(g_1(\alpha))$.
(c) It is shown later that the shape of a pair of centrodes that generate function $g_i(\alpha)$ depends on the derivative $dg_i/d\alpha$. Figures 10.1.3(a), (b), and (c) show the derivatives of transmission functions for the cases wherein the design is based

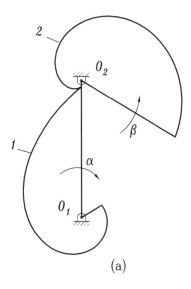

(a)

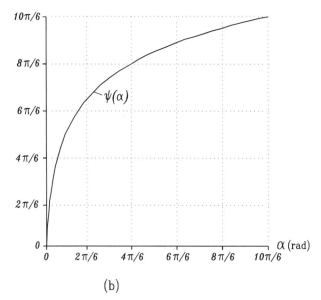

(b)

Figure 10.1.1. (a) Illustration of centrodes of a single pair of noncircular gears; (b) function $\beta(\alpha) \equiv \psi(\alpha)$ generated by centrodes.

on application of (i) one pair of gears (Fig. 10.1.3(a)), (ii) two pairs of gears (Fig. 10.1.3(b)), and (iii) three pairs of gears (Fig. 10.1.3(c)). Centrodes of a multigear drive with three pairs of gears are shown in Fig. 10.1.4.

Drawings of Fig. 10.1.3 confirm that the magnitude of the derivative of the transmission function is substantially decreased in a multigear drive. Another advantage of application of a multigear drive is the increase of the integrated impact of the final output result.

The theory of design of a multigear drive has been developed and is represented here. This chapter also covers the tandem design of a planar linkage coupled with

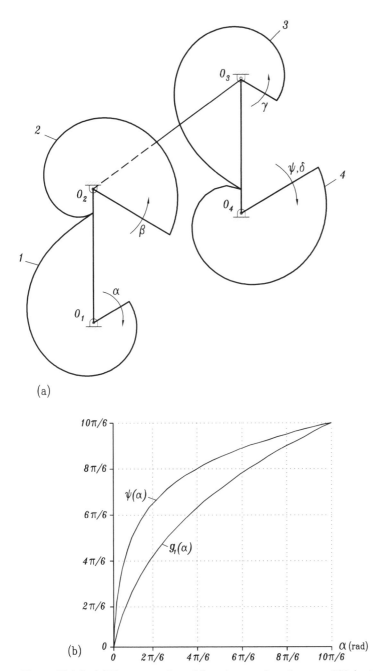

Figure 10.1.2. (a) For generation of function $z = \ln u$, $1 < u < 100$, by two pairs of noncircular gears; (b) function $\psi(\alpha)$ assigned for generation and $g_1(\alpha) \equiv \beta(\alpha)$. Function $\delta(\gamma) \approx \beta(\alpha)$.

noncircular gears for a broader variation of the output speed. This is illustrated with examples of design of:

(a) Tandem design of a double-crank mechanism coupled with two pairs of noncircular gears.

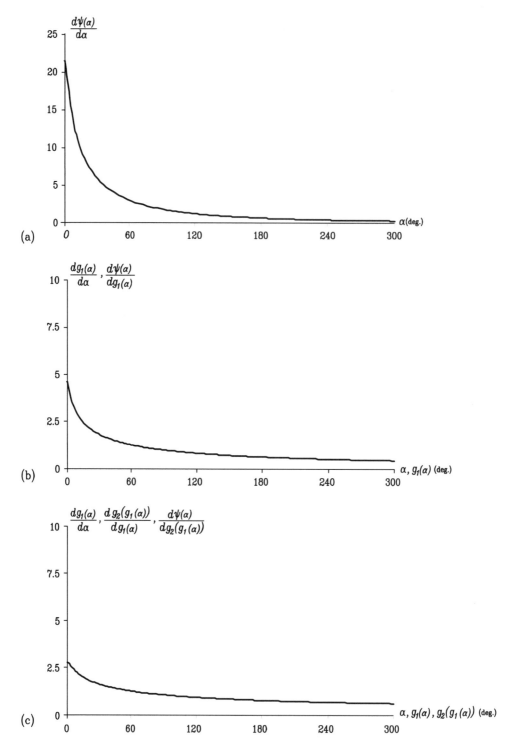

Figure 10.1.3. Illustration of derivative function for (a) a single gear pair; (b) two pairs of gears; (c) three pairs of gears.

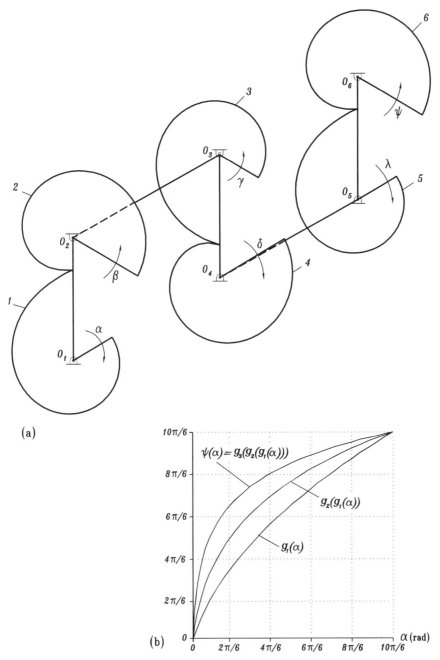

Figure 10.1.4. (a) For generation of function $z = \ln u$, $1 < u < 100$, by three pairs of noncircular gears, (b) transmission functions $g_1(\alpha)$, $g_2(g_1(\alpha))$, and $\psi(\alpha) \equiv g_3(g_2(g_1(\alpha)))$.

(b) Tandem design of a crank-slider linkage coupled with modified elliptical gears.
(c) Tandem design of Scotch-Yoke mechanism coupled with noncircular and circular gears.
(d) Tandem design of mechanism formed by two pairs of noncircular gears and racks.

The developed theory is illustrated with several detailed numerical examples. A functional for generation of a function by a pair of identical centrodes is represented in Section 4.5.

10.2 General Aspects of Generation of Functions

The following conditions must be observed for generation of functions by noncircular gears:

(1) The function chosen for generation must satisfy the requirement for a function to be *strongly monotonic* (Korn & Korn, 1968):

   A real function $f(x)$ of a real variable x is strongly monotonic in (a, b) if $f(x)$ increases as x increases in (a, b) (increasing monotonic function), or if $f(x)$ decreases as x increases in (a, b) (decreasing monotonic function).

(2) We assume that the function to be generated is a strongly increasing monotonic function in the considered interval, and for $x_1 < x_2$ we have that $f(x_1) < f(x_2)$.
(3) Depending on the type of $f(x)$, the centrodes of noncircular gears that generate $f(x)$, might be closed or unclosed curves. The conditions for obtaining centrodes as closed curves are formulated in Section 2.6.
(4) The main idea of generation of $f(x)$ by noncircular gears is that the angles ϕ_1 and ϕ_2 of rotation of gear 1 and 2 are proportional to $(x - x_1)$ and $f(x) - f(x_1)$, respectively.
(5) The scalar coefficients k_1 and k_2 in

$$\phi_1 = k_1(x - x_1), \qquad \phi_2 = k_2[y(x) - y(x_1)] \qquad (10.2.1)$$

are constant and represent the proportionality discussed previously. Coefficients k_1 and k_2 must be assigned observing limitations of pressure angle α_{12} of the gear mechanism (see Section 3.4). In some cases, it may require generation of the function by two (even more) pairs of noncircular gears performing transformation of rotation in sequence (as represented later) for the purpose of reduction of the pressure angle.
(6) It is obvious that although the rotation of noncircular gears is performed with variation of gear ratio, the direction of rotation of the gears cannot be changed in the process of transformation of rotation. Simple derivations yield

$$m_{21}(x) = \frac{d\phi_2}{d\phi_1} = \frac{k_2}{k_1} f'(x) \qquad (10.2.2)$$

It follows from Eq. (10.2.2) that the derivative $f'(x)$ of the function being generated must be positive in the interval of generation, $x_2 \geq x \geq x_1$.
(7) A special approach that allows generation of function $f(x)$ with variation of the sign of the derivative $f'(x)$ is represented in the next section.

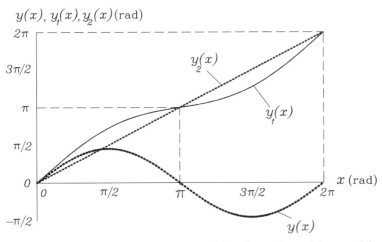

Figure 10.3.1. Illustration of functions $y_1(x) = \sin x + bx$, $y_2(x) = x$, $y(x) = a \sin x$.

10.3 Generation of Function with Varied Sign of Derivative

Function $y(x)$, $x_1 \le x \le x_2$, to be generated by a pair of noncircular gears, must be a monotonous increasing function for which derivative $y'(x) > 0$. The following procedure is applied for generation of a function wherein the derivative $y'(x)$ is of a varied sign in the interval of derivation (Fig. 10.3.1):

(a) Instead of function $y(x)$ assigned for generation (wherein $y'(x)$ is of a varied sign), function

$$y_1(x) = y(x) + bx, \quad x_1 \le x \le x_2, \tag{10.3.1}$$

is generated. The conditions of generation of a function by a pair of noncircular gears, formulated as

$$y_1'(x) \ge 0,$$

may then be observed.

(b) Function $y(x)$ assigned for generation is obtained by the approach illustrated by Fig. 10.3.2 as follows:

(i) Function

$$y_2(x) = bx \tag{10.3.2}$$

is subtracted from $y_1(x)$ by application of a gear differential. Two gear mechanisms formed by (a) noncircular gears 1 and 2, and (b) circular gears 3 and 4, are applied.

(ii) Gears 1 and 3 are mounted on the same shaft, and angles ϕ_1 and ϕ_3 are proportional to the variable x. The design of centrodes of gears 1 and 2 provides that angle of rotation of gear 2 is proportional to the function given by Eq. (10.3.1); similarly, angle of rotation of gear 4 is proportional to the function given by Eq. (10.3.2).

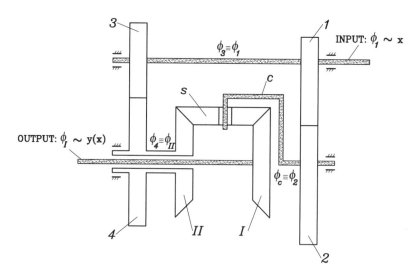

Figure 10.3.2. Structure of gear mechanism formed by a bevel gear differential (bevel gears I and II, satellite s, and carrier c), noncircular gears 1 and 2, and circular gears 3 and 4.

(iii) Rotations of gears 2 and 4 are provided to the carrier c of the satellite s of the gear differential, and to gear II of the differential, respectively.

(iv) The differential provides the following relation between the angles of rotation of the carrier c and gears I and II (Fig. 10.3.2):

$$\phi_I + \phi_{II} = 2\phi_c \qquad (10.3.3)$$

Angle of rotation ϕ_c is equivalent to ϕ_2, and angle of rotation ϕ_{II} is equivalent to ϕ_4.

Equation (10.3.3) yields

$$\phi_I = 2\phi_c - \phi_{II}, \quad \phi_c \equiv \phi_2, \quad \phi_{II} \equiv \phi_4 \qquad (10.3.4)$$

where ϕ_2 is proportional to $y_1(x)$ represented by Eq. (10.3.1) and ϕ_4 is proportional to $y_2(x)$ represented by Eq.(10.3.2).

We assign for the design that

$$\phi_4 = 2b\phi_3, \quad \phi_3 \equiv \phi_1 \quad (y_2(x) = 2x) \qquad (10.3.5)$$

(v) Equations (10.3.4) and (10.3.5) yield that angle ϕ_I of rotation of gear I of the differential is obtained as proportional to function $y(x)$ assigned for design

$$\phi_I = 2\phi_2 - \phi_4 = 2(y(x) + bx) - 2bx = 2y(x) \qquad (10.3.6)$$

Variation of function $y(x)$ by the magnitude and sign will cause variation of angle ϕ_I of rotation of gear I of the differential. This means that gear I will be rotated with varied angular velocity in two directions.

NUMERICAL EXAMPLE 10.3.1: GENERATION OF FUNCTION $y(x) = a\sin x$, $0 \le x \le 2\pi$. The derivative of the to-be-generated function is $y'(x) = a\cos x$ and its sign is varied in the interval of function generation $0 \leq x \leq 2\pi$. The scheme of function generation with varied signed represented by Fig. 10.3.2 is applied.

Noncircular gears 1 and 2 must be designed for generation of function

$$y_1(x) = \sin x + bx, \quad 0 \leq x \leq 2\pi \tag{10.3.7}$$

Design of noncircular gears 1 and 2 (Fig. 10.3.2) must cover determination of their centrodes by application of the following procedure:

(i) The angles of rotation of gears 1 and 2 are represented by

$$\phi_1 = k_1(x - x_1), \quad \phi_2 = k_2[y_1(x) - y_1(x_1)], \quad x_1 = 0 \tag{10.3.8}$$

Here, k_1 and k_2 are scale coefficients determined as

$$k_1 = \frac{\phi_{1,max}}{x_2 - x_1}, \quad k_2 = \frac{\phi_{2,max}}{y_1(x_2) - y_1(x_1)}, \quad x_2 = 2\pi, \quad x_1 = 0 \tag{10.3.9}$$

Taking that gears 1 and 2 will perform in the process of generation turns on $\phi_{1,max} = \phi_{2,max} = 2\pi$, we obtain

$$k_1 = 1, \quad k_2 = \frac{1}{b} \tag{10.3.10}$$

(ii) The coefficient b may be determined by observation of the following conditions: (a) $y'(x) > 0$, and (b) centrodes of gears 1 and 2 must be convex.
(iii) The gear ratio function is

$$m_{21}(x) = \frac{d\phi_2}{d\phi_1} = \frac{\cos x}{b} + 1 \tag{10.3.11}$$

It is obvious that $m_{21}(x) > 0$ because $b > 1$. The conditions of convexity of centrodes 1 and 2 have been represented in Section 2.9 by inequalities represented by Eq. (2.9.2) and (2.9.3). The condition of convexity of centrode 1 yields

$$(\cos x + b)^3 + 2b(1 + b\cos x) + b^2(b + \cos x) \geq 0 \tag{10.3.12}$$

We then assume that there is such a limiting value of b with which the left side of Eq. (10.3.12) does not become equal to zero at $x = \pi$. This yields that the left side of Eq. (10.3.12) is not equal to zero at other values of x of the interval $2\pi \geq x \geq 0$. This allows us to obtain an inequality:

$$(b - 1)^2 - 2b + b^2 \geq 0 \tag{10.3.13}$$

because $b > 1$.

It follows from Eq. (10.3.13) that the limiting value of b is $b_{lim} \geq 1.707$. Analysis of condition of convexity of centrode of gear 2 (centrode 2) yields the same limiting value of b, $b_{lim} \geq 1.707$. Observation of condition $b_{lim} = 1.707$ means that the centrode will have a point with curvature $\kappa_1 = 0$.

(iv) Equations (10.3.8) with coefficients of Eq. (10.3.10) yield the following transmission function,

$$\phi_2(\phi_1) = \phi_1 + \frac{1}{b}\sin\phi_1, \quad 0 \leq \phi_1 \leq 2\pi \tag{10.3.14}$$

The derivative function is

$$m_{12}(\phi_1) = \frac{d\phi_1}{d\phi_2} = \frac{1}{1 + \dfrac{1}{b}\cos\phi_1} \qquad (10.3.15)$$

(v) Equations (2.4.2), (2.4.4), and (2.4.5) (see Chapter 2) yield the following equations for centrodes σ_1 and σ_2 of noncircular gears 1 and 2 (Fig. 10.3.2): For σ_1, we have

$$r_1(\phi_1) = \frac{E}{1 + m_{12}(\phi_1)} = E\frac{1 + \dfrac{1}{b}\cos\phi_1}{2 + \dfrac{1}{b}\cos\phi_1} \qquad (10.3.16)$$

For σ_2, we have

$$r_2(\phi_1) = E\frac{1}{2 + \dfrac{1}{b}\cos\phi_1}, \qquad \phi_2(\phi_1) = \phi_1 + \frac{1}{b}\sin\phi_1 \qquad (10.3.17)$$

Center distance E is just a scale coefficient. Centrodes σ_1 and σ_2, determined with coefficients $b = 1.707$ and $b = 1.400$, are represented in Figs. 10.3.3 and 10.3.4. Centrode 1 in Fig. 10.3.4 is convex-concave.

(vi) We assign for the design that

$$\phi_4 = 2\phi_3, \quad \phi_3 = \phi_1 \qquad (10.3.18)$$

(vii) Function $y(x) = a\sin x$ will be obtained from gear I of the differential as

$$\phi_I = 2\phi_c - \phi_{II} = 2\phi_2 - \phi_4 = 2\left(\phi_1 + \frac{1}{b}\sin\phi_1\right) - 2\phi_1 = \frac{2}{b}\sin\phi_1 \qquad (10.3.19)$$

wherein $\dfrac{2}{b} = a$.

Noncircular gears with the developed centrodes may be generated by the enveloping method by using a hob and a shaper for centrodes shown in Figs. 10.3.3 and 10.3.4, respectively.

10.4 Introduction to Design of Multigear Drive with Noncircular Gears

10.4.1 Basic Functionals

Consider that a given function $\psi(\alpha)$ is generated by a gear drive formed by n pairs of noncircular gears (Fig. 10.4.1). Each pair of gears with conjugated centrodes σ_1 and σ_2 (generally σ_1 does not coincide with σ_2) generates a function $f(\alpha)$. The functional of the gear drive with n pairs is represented as

$$\psi(\alpha) = f(f(f(\ldots f(\alpha)))) = \prod f(\alpha) \qquad (10.4.1)$$

The solution of the functional means determination of function $f(\alpha)$ considering as given $\psi(\alpha)$. Knowing $f(\alpha)$, it becomes possible to derive a pair of centrodes σ_1

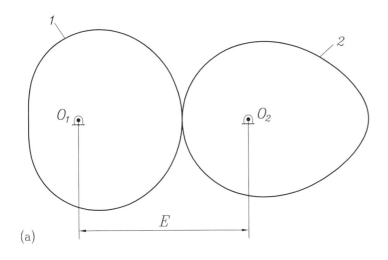

(a)

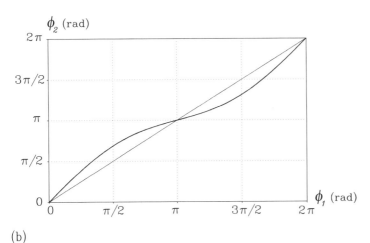

(b)

Figure 10.3.3. Illustration of (a) centrodes of noncircular gears for generation of function $y_1(x)$ with coefficient $b = 1.707$; (b) transmission function $\phi_2(\phi_1)$ wherein coefficient $b = 1.707$.

and σ_2 for generation of function $f(\alpha)$. However, the solution of Functional (10.4.1) (determination of $f(\alpha)$), even if such solution exists, is a difficult problem.

For the case of generation of $\psi(\alpha)$ by application of n pairs of noncircular gears, we use the functional

$$\psi(\alpha) = g_n(g_{n-1}(g_{n-2}(\ldots g_1(\alpha)))), \quad g_k(\alpha) \neq g_{k-1}(\alpha) \qquad (10.4.2)$$

with observation of following conditions:

(i) Each function g_k ($k = 1, \ldots, n$) is generated by a pair of conjugated centrodes $(\sigma_1^{(k)}, \sigma_2^{(k)})$, $k = 1, \ldots, n$.

(ii) Centrode $\sigma_1^{(k)}$ is mounted on the shaft of centrode $\sigma_2^{(k-1)}$, and the angles of rotation ϕ_{2k-1} and ϕ_{2k-2} of centrodes $\sigma_1^{(k)}$ and $\sigma_2^{(k-1)}$, respectively, are equal (see Fig. 10.4.2).

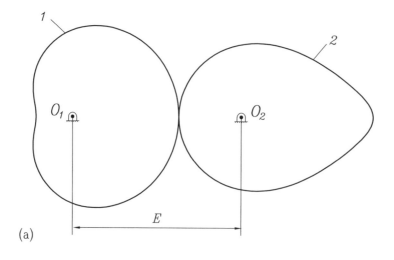

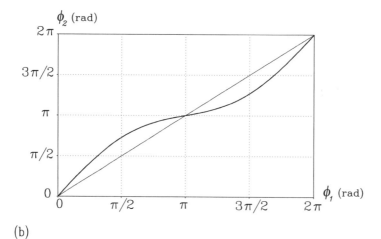

(b)

Figure 10.3.4. Illustration of (a) centrodes of noncircular gears for generation of function $y_1(x)$ with coefficient $b = 1.400$; (b) transmission function $\phi_2(\phi_1)$ wherein coefficient $b = 1.400$.

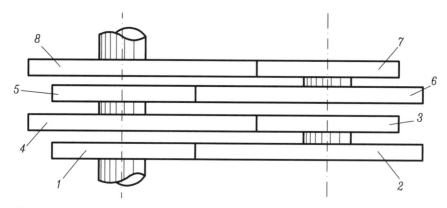

Figure 10.4.1. Schematic illustration of assembly of four pairs of gears on collinear shafts.

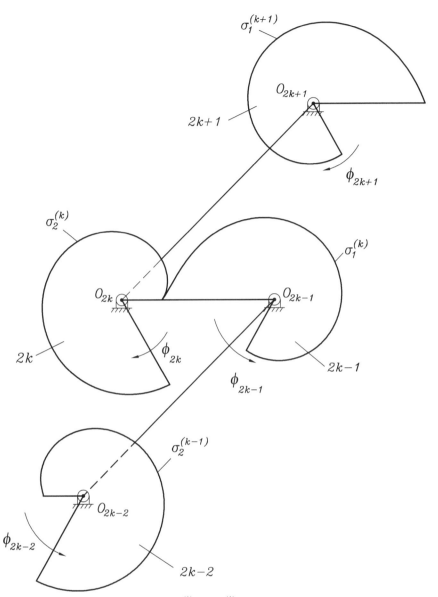

Figure 10.4.2. A pair of centrodes $\sigma_1^{(k)}$ and $\sigma_2^{(k)}$ in a gear drive formed by n pairs of centrodes.

(iii) Angle α is the input parameter and represents the angle of rotation of centrode $\sigma_1^{(1)}$ of the pair of centrodes $(\sigma_1^{(1)}, \sigma_2^{(1)})$. The angle of rotation of centrode $\sigma_1^{(k)}$ is proportional to function $g_{k-1}(\alpha) \equiv \phi_{2k-1}(\alpha)$ (see Fig. 10.4.2). The angle of rotation of centrode $\sigma_2^{(k)}$ is proportional to function $g_k(\alpha) \equiv \phi_{2k}(\alpha)$ (see Fig. 10.4.2). The angle of rotation of centrode $\sigma_2^{(n)}$ is proportional to the generated function $\psi(\alpha)$.

(iv) The maximal angles of rotation of any pair of conjugated centrodes $(\sigma_1^{(k)}, \sigma_2^{(k)})$ are equal. The generation of all functions $g_n(g_{n-1}(\cdots g_2(g_1(\alpha))))$ is started and finished at the same positions of the gear drive formed by noncircular gears.

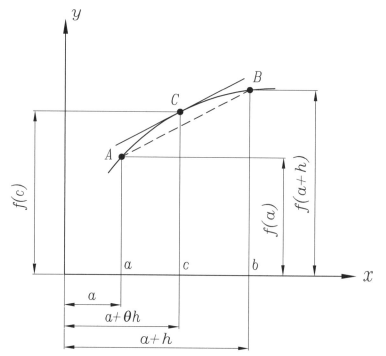

Figure 10.4.3. Illustration of mean value function and function increase.

(v) The developed algorithm for generation of given function $\psi(\alpha)$ is based on derivation of centrodes wherein $g_k(\alpha) \neq g_{k-1}(\alpha)$. The generation of given function $\psi(\alpha)$ by the multigear drive is performed exactly.

(vi) Generation of $\psi(\alpha)$ by Functional (10.4.2) is in some sense equivalent to the split of $\psi(\alpha)$ between the generating gears (centrodes). It allows us to reduce the pressure angle and provide a favorable shape for the centrodes. This great advantage is achieved due to the reduction of the derivatives of functions $g_k(g_{k-1})$ with respect to function $\psi(\alpha)$ (see numerical examples following).

(vii) The split of function $\psi(\alpha)$ by its generation (see Functional (10.4.2)) may be applied as well for technological processes, not only for generation of functions.

10.4.2 Interpretations of Lagrange's Theorem

Two interpretations of Lagrange's theorem are considered:

(a) **Mean Value Theorem** (Korn & Korn, 1968). Assume that function $f(x)$ is continuous on the closed interval $[a, b]$ and continuously differentiable on the open interval (a, b). There is then such a point c (Fig. 10.4.3) where

$$\frac{f(b) - f(a)}{b - a} = f'(c), \quad x(c) = a + \theta(b - a), \quad 0 < \theta < 1 \qquad (10.4.3)$$

Equation (10.4.3) means that on the smooth curve (Fig. 10.4.3) between (A, B), there is such a point C at which the tangent is parallel to the chord $A - B$.

(b) **Final Increment of Function** (Fichtengolz, 1958). Using Eq. (10.4.3), we may formulate the final increase of a function. Assume that at point $x = x_o$, there is an increase of Δx. Equation (10.4.3) then yields

$$\frac{f(x_o + \Delta x) - f(x_o)}{\Delta x} = f'(x_o + \theta \Delta x), \quad 0 < \theta < 1 \tag{10.4.4}$$

The increase $\Delta f(x_o)$ of the function is represented as $\Delta f(x_o) = f(x_o + \Delta x) - f(x_o)$.

Assume now that we consider the increase of function $f(a)$ at point a, where the increase Δx is $h = b - a$; then we obtain from Eq. (10.4.4) that the increase of function on the interval (a, b) is

$$\frac{f(a + h) - f(a)}{h} = f'(a + \theta h), \quad 0 < \theta < 1 \tag{10.4.5}$$

10.4.3 Illustration of Application of Lagrange's Theorem for Functional $\psi(\alpha) = g_2(g_1(\alpha))$

The functional applied for generation of given function $\psi(\alpha)$ is represented as

$$\psi(\alpha) = g_2(g_1(\alpha)) \tag{10.4.6}$$

Centrodes $\sigma_1^{(1)}$ and $\sigma_2^{(1)}$ (see Fig. 10.1.2(a)) generate function $g_1(\alpha) \equiv \beta(\alpha)$. Centrodes $\sigma_1^{(2)}$ and $\sigma_2^{(2)}$ (see Fig. 10.1.2(a)) generate function $g_2(\alpha) \equiv \delta(\alpha)$. Function $g_2(\alpha) \neq g_1(\alpha)$ and centrodes $\sigma_1^{(1)}$ and $\sigma_2^{(1)}$ are not identical to centrodes $\sigma_1^{(2)}$ and $\sigma_2^{(2)}$, respectively. The centrodes might be as well non-closed, but the maximal angles of rotation of all centrodes must be equal.

10.4.3.1 Previous Solutions for $\psi(\alpha) = f(f(\alpha))$.

Collatz approach (Collatz, 1951) is based on the following considerations:

(i) It is assumed that the solution for the functional

$$\psi(\alpha) = f(f(\alpha)) \tag{10.4.7}$$

exists.

(ii) In addition, it is assumed that function $\psi(\alpha)$ may be represented as a power series

$$\psi(\alpha) = \sum_{n=1}^{\infty} \psi_n \alpha^n \tag{10.4.8}$$

(iii) The solution of Functional (10.4.7) may then be obtained as

$$f(\alpha) = \sum_{n=0}^{\infty} a_n \alpha^n \tag{10.4.9}$$

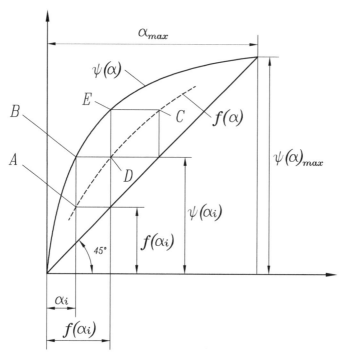

Figure 10.4.4. Illustration of discrete representation of function $f(\alpha)$ as the solution of functional $\psi(\alpha) = f(f(\alpha))$ with given function $\psi(\alpha)$.

The coefficients a_n are determined by solution of nonlinear equations obtained by simultaneous consideration of Eqs. (10.4.7), (10.4.8), and (10.4.9). A computerized procedure for determination of coefficients a_n in Eq. (10.4.9) is to be provided.

Collatz has considered as well a particular case wherein the sought-for function $f(\alpha)$ may be represented discretely. Consider functional $\psi(\alpha) = f(f(\alpha))$ with a given function $\psi(\alpha)$. It is known from Collatz (Collatz, 1951) that $f(\alpha)$ may be represented discretely if a single point A of $f(\alpha)$ is known exactly, and the limitation is that $f(\alpha_i) \neq i$.

Drawings of Fig. 10.4.4 show the discrete presentation of $f(\alpha)$ that is based on the relations

$$\alpha_{i+1} = f(\alpha_i), \qquad f(\alpha_{i+1}) = \psi(\alpha_i) \qquad (10.4.10)$$

Considering point A of $f(\alpha_i)$ as known, points D, C, ..., of $f(\alpha)$ may be determined.

Differentiation of Eq. (10.4.10) provides

$$f'(\alpha_i) \cdot f'(\alpha_{i+1}) = \psi'(\alpha_i) \qquad (10.4.11)$$

Denoting α_i and α_{i+1} as conjugated, we may represent Eq. (10.4.11) as

$$f'(\alpha_i) \cdot f'[f(\alpha_i)] = \psi'(\alpha_i) \qquad (10.4.12)$$

The disadvantages of the discrete presentation of $f(\alpha)$ are (i) the exact solution of $\psi(\alpha) = f(f(\alpha))$ is required for at least one point of $f(\alpha)$; (ii) the location of conjugated points of $f(\alpha)$ is prescribed ahead. However, an iterative process for discrete presentation of $f(\alpha)$ may be applied.

An approximate solution of Functional (10.4.7) was proposed by Kislitzin (Kislitzin, 1955) by considering the functional

$$\phi(\phi(x)) = f(x) \tag{10.4.13}$$

The approximate solution of Functional (10.4.13) was represented by Kislitzin as

$$\phi(x) \approx \frac{f(x) + x\sqrt{f'(x)}}{1 + \sqrt{f'(x)}} \tag{10.4.14}$$

Litvin (Litvin, 1956) has considered the Functional (10.4.6) and applied the Lagrange's theorem for the mean value of a function (Korn & Korn, 1968) for derivation of $g_1(\alpha)$ (see below). In addition, Litvin has proposed the derivation of centrodes for $g_1(\alpha)$ and $g_2(\alpha)$ that provide exact generation of $\psi(\alpha)$.

10.4.3.2 Computational Procedure for Functional (10.4.6).
The approach proposed by Litvin (Litvin, 1956) has been extended now for gear drives with n pairs of centrodes.

The approach for a gear drive with two pairs of gears is formulated as follows:

(1) Lagrange equation (similar to Eq. (10.4.5)) states

$$\frac{g_1(\alpha + h) - g_1(\alpha)}{h} = g_1'(\alpha + \theta h), \quad 0 < \theta < 1 \tag{10.4.15}$$

(2) Taking in Eq. (10.4.15) $h = g_1(\alpha) - \alpha, h \neq 0$, we obtain

$$g_1(\alpha + h) = g_1\{\alpha + [g_1(\alpha) - \alpha]\} = g_1(g_1(\alpha)) = \psi(\alpha) \tag{10.4.16}$$

(3) Equations (10.4.15) and (10.4.16) yield

$$\frac{\psi(\alpha) - g_1(\alpha)}{g_1(\alpha) - \alpha} = g_1'\{\alpha + \theta[g_1(\alpha) - \alpha]\} \tag{10.4.17}$$

(4) It follows from Eq. (10.4.16) that

$$g_1'(g_1(\alpha))g_1'(\alpha) = \psi'(\alpha) \tag{10.4.18}$$

(5) The sought-for function $g_1(\alpha)$ should not deviate substantially from a straight line at the interval h, and therefore we may take that

$$g_1'(\alpha) \approx g_1'[g_1(\alpha)] \approx g_1'\{\alpha + \theta[g_1(\alpha) - \alpha]\} \tag{10.4.19}$$

(6) Equations (10.4.18) and (10.4.19) yield

$$g_1'(\alpha) \approx g_1'[g_1(\alpha)] \approx g_1'\{\alpha + \theta[g_1(\alpha) - \alpha]\} \approx \sqrt{\psi'(\alpha)} \tag{10.4.20}$$

(7) Considering Eqs. (10.4.17) and (10.4.20), we obtain

$$\frac{\psi(\alpha) - g_1(\alpha)}{g_1(\alpha) - \alpha} \approx \sqrt{\psi'(\alpha)} \qquad (10.4.21)$$

(8) Solution of Eq. (10.4.21) for $g_1(\alpha)$ yields the following expression of $g_1(\alpha)$:

$$g_1(\alpha) \approx \frac{\psi(\alpha) + \alpha\sqrt{\psi'(\alpha)}}{1 + \sqrt{\psi'(\alpha)}} \qquad (10.4.22)$$

(9) Equation (10.4.22) is the approximate solution for functional

$$\psi(\alpha) = g_2(g_1(\alpha)) \qquad (10.4.23)$$

The importance of Eq. (10.4.22) is that knowing $\psi(\alpha)$, α, and $\psi'(\alpha)$, it becomes possible to derive the coupled noncircular gears (Fig. 10.1.2).

The procedure of computation of centrodes provides that centrodes $(\sigma_1^{(1)}, \sigma_1^{(2)})$ and $(\sigma_2^{(1)}, \sigma_2^{(2)})$ are not identical. A procedure of derivation of coupled centrodes is proposed that allows exact generation of function $\psi(\alpha)$.

NUMERICAL EXAMPLE 10.4.1: GENERATION OF FUNCTION $z = \ln u$, $1 < u < 100$, BY A GEAR DRIVE WITH $n = 1$ PAIR OF NONCIRCULAR GEARS. A gear drive based on one pair of noncircular gears (Fig. 10.1.1) is applied for function generation. Derivation of centrodes σ_1 and σ_2 is based on the following procedure:

(1) Intervals of rotation of the two centrodes are considered to be equal:

$$0 < \alpha < 10\pi/6, \quad 0 < \psi < 10\pi/6$$

(2) Angle of rotation of centrode $\sigma_1^{(1)}$, α, is proportional to independent variable u, and scalar coefficient k_1 is determined from

$$\alpha = k_1(u - u_{min}) + \alpha_{min}$$

Here, $u_{min} = 1$, $\alpha_{min} = 0$ rad.

(3) Angle of rotation of centrode $\sigma_2^{(2)}$, ψ, is proportional to dependent variable z, and scalar coefficient k_2 is determined from

$$\psi = k_2(z - z_{min}) + \psi_{min}$$

Here, $z_{min} = \ln 1$, $\psi_{min} = 0$ rad.

(4) Scalar coefficients k_1 and k_2 are obtained as

$$k_1 = \frac{\alpha_{max} - \alpha_{min}}{u_{max} - u_{min}} = \frac{10\pi/6 - 0}{100 - 1}$$

$$k_2 = \frac{\psi_{max} - \psi_{min}}{z_{max} - z_{min}} = \frac{10\pi/6 - 0}{\ln 100 - \ln 1}$$

(5) Function $\psi(\alpha)$ is represented as

$$\psi(\alpha) = k_2 \left[\ln \left(\frac{\alpha - \alpha_{min}}{k_1} + u_{min} \right) - z_{min} \right] + \psi_{min}$$

(6) Function $\psi(\alpha)$ allows us to determine the centrodes σ_1 and σ_2 in their respective coordinate systems S_1 and S_2 as

$$r_1 = E\frac{1}{1+m_{12}} = E\frac{1}{1+\dfrac{1}{\psi'(\alpha)}}$$

$$r_2 = E - r_1^{(1)}$$

$$r_{1x} = -r_1^{(1)}\sin(\alpha - \alpha_{min})$$

$$r_{1y} = +r_1^{(1)}\cos(\alpha - \alpha_{min})$$

$$r_{2x} = -r_2^{(1)}\sin(\psi - \psi_{min})$$

$$r_{2y} = -r_2^{(1)}\cos(\psi - \psi_{min})$$

Here, E is the center distance chosen in this numerical example as $E = 400$ mm.

Figure 10.1.1(a) shows the pair of centrodes, and Figure 10.1.1(b) shows the generated function $\psi(\alpha)$.

NUMERICAL EXAMPLE 10.4.2: GENERATION OF FUNCTION $z = \ln u$, $1 < u < 100$, BY A GEAR DRIVE WITH $n = 2$ PAIRS OF NONCIRCULAR GEARS. A gear drive based on two pairs of noncircular gears with center distance $E = 400$ mm is applied for the purpose of function generation (see Fig. 10.1.2). Derivation of centrodes $\sigma_1^{(1)}, \sigma_2^{(1)}, \sigma_1^{(2)}, \sigma_2^{(2)}$ is based on the following procedure:

(1) Intervals of rotation of all centrodes are considered as equal:

$$0 < \alpha < 10\pi/6, \quad 0 < \beta < 10\pi/6, \quad 0 < \gamma < 10\pi/6, \quad 0 < \psi < 10\pi/6$$

(2) Angle of rotation of centrode $\sigma_1^{(1)}$, α, is proportional to independent variable u, and the scalar coefficient k_1 is obtained from

$$\alpha = k_1(u - u_{min}) + \alpha_{min}$$

Here, $u_{min} = 1$, $\alpha_{min} = 0$ rad.

(3) Angle of rotation of centrode $\sigma_2^{(2)}$, ψ, is proportional to dependent variable z, and the scalar coefficient k_2 is obtained from

$$\psi = k_2(z - z_{min}) + \psi_{min}$$

Here, $z_{min} = \ln 1$, $\psi_{min} = 0$ rad.

(4) We then have

$$k_1 = \frac{\alpha_{max} - \alpha_{min}}{u_{max} - u_{min}} = \frac{10\pi/6 - 0}{100 - 1}$$

$$k_2 = \frac{\psi_{max} - \psi_{min}}{z_{max} - z_{min}} = \frac{10\pi/6 - 0}{\ln 100 - \ln 1}$$

(5) Function $\psi(\alpha)$ is derived as

$$\psi(\alpha) = k_2 \left[\ln \left(\frac{\alpha - \alpha_{min}}{k_1} + u_{min} \right) - z_{min} \right] + \psi_{min}$$

(6) Function $\beta(\alpha) \equiv g_1(\alpha)$ generated by the first pair of centrodes is represented by Eq. (10.4.22).

(7) Function $g_1(\alpha)$ allows us to determine the centrodes $\sigma_1^{(1)}$ and $\sigma_2^{(1)}$ at their respective coordinate systems S_1 and S_2 as

$$r_1 = E \frac{1}{1 + m_{12}} = E \frac{1}{1 + \dfrac{1}{g_1'(\alpha)}}$$

$$r_2 = E - r_1^{(1)}$$

$$r_{1x} = -r_1 \sin(\alpha - \alpha_{min})$$

$$r_{1y} = +r_1 \cos(\alpha - \alpha_{min})$$

$$r_{2x} = -r_2 \sin(\beta - \beta_{min})$$

$$r_{2y} = -r_2 \cos(\beta - \beta_{min})$$

(8) Functions $\psi(\alpha)$ and $g_1(\alpha)$ allow us to determine the centrodes $\sigma_1^{(2)}$ and $\sigma_2^{(2)}$ at their respective coordinate systems S_3 and S_4 as

$$r_3 = E \frac{1}{1 + m_{34}} = E \frac{1}{1 + \dfrac{g_1'(\alpha)}{\psi'(\alpha)}}$$

$$r_4 = E - r_1^{(2)}$$

$$r_{3x} = -r_3 \sin(\beta - \beta_{min})$$

$$r_{3y} = -r_3 \cos(\beta - \beta_{min})$$

$$r_{4x} = -r_4 \sin(\psi - \psi_{min})$$

$$r_{4y} = +r_4 \cos(\psi - \psi_{min})$$

Figure 10.1.2(a) shows the two pairs of centrodes, and Fig. 10.1.2(b) shows the generated functions $g_1(\alpha)$ and $\psi(\alpha)$.

10.5 Design of Multigear Drive

10.5.1 Basic Equations

A multigear drive based on n pairs of noncircular gears must be designed for generation of a given function $\psi(\alpha)$. Here, α is the angle of rotation of centrode 1, of pair 1, and ψ is the angle of rotation of driven centrode of pair n.

Functional (10.4.2) is used for generation of the given function $\psi(\alpha)$. The goal of this section is determination of functions $g_k(\alpha)$ (see Eq. (10.4.2)) generated by each pair k, $k = 1, \ldots, n$, in the form

$$g_k = g_k(g_{k-1}, \psi(\alpha), \psi'(\alpha)) \tag{10.5.1}$$

Here, g_k is the angle of rotation of driven centrode $2k$ and g_{k-1} is the angle of rotation of centrode $2k - 1$ (the driving one). In the case that $k = 1$, $g_{k-1} = g_0 = \alpha$ is the angle of rotation of centrode 1.

The procedure is as follows:

(1) Differentiation of Functional (10.4.2) provides that

$$\psi'(\alpha) = g_n'(g_{n-1}) \cdot g_{n-1}'(g_{n-2}) \cdot \ldots \cdot g_2'(g_1) \cdot g_1'(\alpha) \tag{10.5.2}$$

(2) We may assume that $g_n'(g_{n-1}) \approx g_{n-1}'(g_{n-2}) \approx \ldots \approx g_2'(g_1) \approx g_1'(\alpha)$. This assumption is based on the tendency of function g_k toward a straight line when n trends toward infinity. Then the following relations are observed:

$$g_k'(g_{k-1}) \approx [\psi'(\alpha)]^{1/n}, \quad k = 1, \ldots, n \tag{10.5.3}$$

(3) Considering now Lagrange's theorem for each function $g_k(g_{k-1})$, $k = 1, \ldots, n$,

$$g_k'(g_{k-1} + \theta h_k) = \frac{g_k(g_{k-1} + h_k) - g_k(g_{k-1})}{h_k}, \quad 0 < \theta < 1 \tag{10.5.4}$$

and taking $h_k = g_k(g_{k-1}) - g_{k-1}$, we obtain

$$g_k'(g_{k-1} + \theta h_k) = \frac{g_k(g_{k-1} + g_k(g_{k-1}) - g_{k-1}) - g_k(g_{k-1})}{g_k(g_{k-1}) - g_{k-1}} \tag{10.5.5}$$

(4) Considering Eqs. (10.5.3) and (10.5.5), we may determine g_k as a function of g_{k-1}, g_{k+1}, and $\psi'(\alpha)$

$$\frac{g_k(g_k(g_{k-1})) - g_k(g_{k-1})}{g_k(g_{k-1}) - g_{k-1}} \approx [\psi'(\alpha)]^{1/n} \tag{10.5.6}$$

In this equation, $g_k(g_k(g_{k-1})) \approx g_{k+1}(g_k)$ so that

$$\frac{g_{k+1}(g_k) - g_k(g_{k-1})}{g_k(g_{k-1}) - g_{k-1}} \approx [\psi'(\alpha)]^{1/n} \tag{10.5.7}$$

and finally we can determine $g_k(g_{k-1})$ as

$$g_k(g_{k-1}) \approx \frac{g_{k+1}(g_k) + g_{k-1}[\psi'(\alpha)]^{1/n}}{1 + [\psi'(\alpha)]^{1/n}} \tag{10.5.8}$$

(5) By observing Eq. (10.5.8) from $k = 1$ to $k = n - 1$ and considering that $g_n = \psi(\alpha)$ and $g_0 = \alpha$, the following general equation is determined:

$$g_k(\alpha) \approx \frac{\psi(\alpha) + g_{k-1}\left[\sum_{j=1}^{j=n-k}[\psi'(\alpha)]^{j/n}\right]}{1 + \sum_{j=1}^{j=n-k}[\psi'(\alpha)]^{j/n}}, \quad k = 1, \ldots, n - 1 \tag{10.5.9}$$

10.5.2 Design of Centrodes

Design of the centrodes of noncircular gears requires determination of the gear ratio for each pair k $(k = 1, \ldots, n)$ as

$$m_{2k-1,2k} = \frac{dg_{k-1}}{dg_k} = \frac{g'_{k-1}(\alpha)}{g'_k(\alpha)}, \quad k = 1, \ldots, n \qquad (10.5.10)$$

Polar vectors are then obtained as

$$r_{2k-1} = E \frac{1}{1 + m_{2k-1,2k}}, \quad k = 1, \ldots, n \qquad (10.5.11)$$

$$r_{2k} = E - r_{2k-1}, \quad k = 1, \ldots, n \qquad (10.5.12)$$

Here, E is the center distance.

NUMERICAL EXAMPLE 10.5.1: GENERATION OF FUNCTION $z = \ln u$, $1 < u < 100$ BY A GEAR DRIVE WITH $n = 3$ PAIRS OF NONCIRCULAR GEARS. The procedure for determination of the six centrodes is as follows:

(1) The functional

$$\psi(\alpha) = g_3(g_2(g_1(\alpha))) \qquad (10.5.13)$$

is formed by three functions generated by three pairs of centrodes. Centrodes $\sigma_1^{(1)}$ and $\sigma_2^{(1)}$ provide function $g_1(\alpha)$, centrodes $\sigma_1^{(2)}$ and $\sigma_2^{(2)}$ provide function $g_2(\alpha)$, and centrodes $\sigma_1^{(3)}$ and $\sigma_2^{(3)}$ provide function $g_3(\alpha)$ (see Fig. 10.5.1). The centrodes are unclosed curves and the maximal angles of rotation of all six centrodes are equal and of $10\pi/6$ radians.

(2) Angle of rotation of centrode $\sigma_1^{(1)}$ is proportional to variable u, whereas angle of rotation of centrode $\sigma_2^{(3)}$ is proportional to variable z:

$$\alpha = k_1(u - u_{min}) + \alpha_{min} = k_1(u - 1) \qquad (10.5.14)$$

$$\psi = k_2(\ln u - \ln u_{min}) + \psi_{min} = k_2 \ln u \qquad (10.5.15)$$

Here, $\alpha_{min} = 0$ and $\psi_{min} = 0$.

(3) Coefficients k_1 and k_2 are determined as

$$k_1 = \frac{\alpha_{max} - \alpha_{min}}{u_{max} - u_{min}} = \frac{10\pi/6}{100 - 1}$$

$$k_2 = \frac{\psi_{max} - \psi_{min}}{\ln u_{max} - \ln u_{min}} = \frac{10\pi/6}{\ln 100 - \ln 1}$$

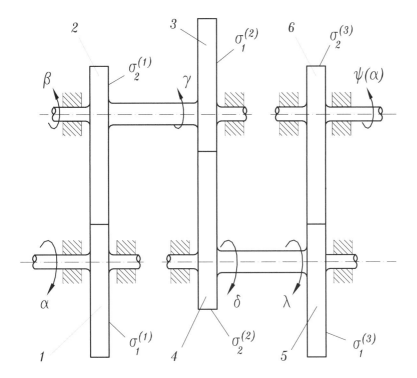

Figure 10.5.1. Schematic illustration of generation of function $\psi(\alpha) = g_3(g_2(g_1(\alpha)))$ by three pairs of noncircular gears that generate respectively $\beta = g_1(\alpha)$, $\delta = g_2(\alpha)$, and $\psi = g_3(\alpha)$.

(4) Function $\psi(\alpha)$ is then derived as

$$\psi(\alpha) = k_2 \left[\ln\left(\frac{\alpha}{k_1} + 1 \right) - \ln 1 \right] \tag{10.5.16}$$

(5) Function $g_1(\alpha) \equiv \beta$ is determined by application of Eq. (10.5.9) with $k = 1$ and $n = 3$,

$$g_1(\alpha) = \frac{\psi(\alpha) + \alpha[\psi'(\alpha)]^{1/3} + \alpha[\psi'(\alpha)]^{2/3}}{1 + [\psi'(\alpha)]^{1/3} + [\psi'(\alpha)]^{2/3}} \tag{10.5.17}$$

wherein

$$\psi'(\alpha) = \frac{k_2}{\alpha + k_1} \tag{10.5.18}$$

(6) Function $g_2(\alpha) \equiv \delta$ is determined by application of Eq. (10.5.9) with $k = 2$ and $n = 3$,

$$g_2(\alpha) = \frac{\psi(\alpha) + g_1(\alpha)[\psi'(\alpha)]^{1/3}}{1 + [\psi'(\alpha)]^{1/3}} \tag{10.5.19}$$

(7) Centrodes $\sigma_1^{(1)}$ and $\sigma_2^{(1)}$ are determined in their coordinate systems S_1 and S_2 as

$$r_1(\alpha) = E\frac{1}{1+m_{12}(\alpha)} = E\frac{1}{1+\dfrac{1}{g_1'(\alpha)}}$$

$$r_2(g_1(\alpha)) = E - r_1(\alpha)$$

$$r_{1x} = -r_1 \sin(\alpha - \alpha_{min})$$

$$r_{1y} = +r_1 \cos(\alpha - \alpha_{min})$$

$$r_{2x} = -r_2 \sin(g_1(\alpha) - g_1(\alpha_{min}))$$

$$r_{2y} = -r_2 \cos(g_1(\alpha) - g_1(\alpha_{min}))$$

(8) Centrodes $\sigma_1^{(2)}$ and $\sigma_2^{(2)}$ are determined in their coordinate systems S_3 and S_4 as

$$r_3(g_1(\alpha)) = E\frac{1}{1+m_{34}(g_1(\alpha))} = E\frac{1}{1+\dfrac{g_1'(\alpha)}{g_2'(\alpha)}}$$

$$r_4(g_2(g_1(\alpha))) = E - r_3(g_1(\alpha))$$

(10.5.20)

$$r_{3x} = -r_3 \sin(g_1(\alpha) - g_1(\alpha_{min}))$$

$$r_{3y} = -r_3 \cos(g_1(\alpha) - g_1(\alpha_{min}))$$

$$r_{4x} = -r_4 \sin(g_2(g_1(\alpha)) - g_2(g_1(\alpha_{min})))$$

$$r_{4y} = +r_4 \cos(g_2(g_1(\alpha)) - g_2(g_1(\alpha_{min})))$$

(9) Centrodes $\sigma_1^{(3)}$ and $\sigma_2^{(3)}$ are determined in their coordinate systems S_5 and S_6 as

$$r_5(g_2(g_1(\alpha))) = E\frac{1}{1+m_{56}(g_2(g_1(\alpha)))} = E\frac{1}{1+\dfrac{g_2'(\alpha)}{\psi'(\alpha)}}$$

$$r_6(\psi) = E - r_5(g_2(g_1(\alpha)))$$

$$r_{5x} = -r_5 \sin(g_2(g_1(\alpha)) - g_2(g_1(\alpha_{min})))$$

$$r_{5y} = +r_5 \cos(g_2(g_1(\alpha)) - g_2(g_1(\alpha_{min})))$$

$$r_{6x} = -r_6 \sin(\psi(\alpha) - \psi(\alpha_{min}))$$

$$r_{6y} = -r_6 \cos(\psi(\alpha) - \psi(\alpha_{min}))$$

Figure 10.1.4(b) shows the generated functions $g_1(\alpha) \equiv \beta(\alpha)$, $g_2(\alpha) \equiv \delta(\alpha)$, and $g_3(\alpha) \equiv \psi(\alpha)$.

Figures 10.1.1, 10.1.2, and 10.1.4, show the influence of the number of the applied gear drives on the shape of centrodes. The main factor of the influence is the derivative of transmission function (Fig. 10.1.3).

10.6 Design of Planar Linkages Coupled with Noncircular Gears

This section covers the following examples of synthesis of mechanisms:

 (i) Generation of a function by a double-crank linkage coupled with two pairs of noncircular gears (Section 10.6.1).
 (ii) Tandem design of a crank-slider linkage coupled with modified elliptical gears (Section 10.6.2).
(iii) Tandem design of a Scotch-Yoke mechanism coupled with noncircular and circular gears (Section 10.6.3).
(iv) Tandem design of mechanism formed by two pairs of noncircular gears and racks (Section 10.6.4).

The combination of a planar linkage with noncircular gears is very useful for variation of output speed variation (Sections 10.6.2 and 10.6.3). However, in the case of generation of functions, the preferable design should be based on application of multigear drives formed by noncircular gears, but not a tandem design of a planar linkage with noncircular gears.

The disadvantage of the previously mentioned case (i) (wherein a planar linkage is coupled with noncircular gears for function generation) is the result of the substantial difference of the number of design parameters with respect to the design of multigear drives formed by noncircular gears, wherein the number of design parameter is much greater.

Apart of mechanisms represented in Sections 10.6.1 to 10.6.3, the mechanism formed by two pairs of racks and noncircular gears (Section 10.6.4) can be applied for variation of output speed and function generation. Application of two pairs of racks and noncircular gear allows obtaining larger magnitude of translational motion.

10.6.1 Tandem Design of Double-Crank Mechanism Coupled with Two Pairs of Noncircular Gears

The limited number of design parameters of a four-bar linkage causes a handicap when a given function is required to be provided. In this sense, tandem design of the linkage with noncircular gears is helpful for reducing of the differences between the desired function and the governing function of the linkage (see numerical example later).

NUMERICAL EXAMPLE 10.6.1: APPROXIMATE GENERATION OF FUNCTION $z = \ln u$, $1 < u <$ 10, BY A FOUR-BAR LINKAGE AND TWO PAIRS OF NONCIRCULAR GEARS. Function $z = \ln u$, $1 < u < 10$, is considered to be obtained through a tandem design based on two pairs of noncircular gears and a four-bar double-crank linkage. The applied functional is represented by $\psi(\alpha) = g_3(g_2(g_1(\alpha)))$, but the functional to be obtained is $\psi^*(\alpha) = g_3^*(g_2(g_1(\alpha)))$, wherein function g_3^* must be generated by a four-bar linkage instead of function g_3, and functions g_1 and g_2 will be generated by two pairs of noncircular gears.

Rotation α is proportional to independent variable u by application of coefficient k_1:

$$\alpha = k_1(u - u_{min}) + \alpha_{min} \tag{10.6.1}$$

Rotation ψ is proportional to dependent variable z by application of coefficient k_2:

$$\psi = k_2(z(u) - z(u_{min})) + \psi_{min} \tag{10.6.2}$$

Function $\psi(\alpha)$ is then derived as

$$\psi(\alpha) = k_2 \left[\ln \left(\frac{\alpha - \alpha_{min}}{k_1} + u_{min} \right) - \ln u_{min} \right] + \psi_{min} \tag{10.6.3}$$

The intervals of rotations are chosen as

$$\alpha_{min} = 0; \quad \alpha_{max} = 11\pi/6; \quad \psi_{min} = -2\pi/6; \quad \psi_{max} = 8\pi/6$$

and provide coefficients k_1 and k_2:

$$k_1 = \frac{\alpha_{max} - \alpha_{min}}{u_{max} - u_{min}} = \frac{11\pi/6 - 0}{10 - 1}$$

$$k_2 = \frac{\psi_{max} - \psi_{min}}{z_{max} - z_{min}} = \frac{8\pi/6 - (-2\pi/6)}{\ln 10 - \ln 1}$$

Function $g_1(\alpha) \equiv \beta(\alpha)$ is derived as

$$g_1(\alpha) \equiv \beta(\alpha) = \frac{\psi(\alpha) + \alpha[\psi'(\alpha)]^{1/3} + \alpha[\psi'(\alpha)]^{2/3}}{1 + [\psi'(\alpha)]^{1/3} + [\psi'(\alpha)]^{2/3}} \tag{10.6.4}$$

wherein $\psi'(\alpha)$ is given as

$$\psi'(\alpha) = \frac{k_2}{\alpha - \alpha_{min} + u_{min}k_1} \tag{10.6.5}$$

Function $g_2(\alpha) \equiv \delta(\alpha)$ is given as

$$g_2(\alpha) \equiv \delta(\alpha) = \frac{\psi(\alpha) + g_1(\alpha)[\psi'(\alpha)]^{1/3}}{1 + [\psi'(\alpha)]^{1/3}} \tag{10.6.6}$$

For the synthesis of the linkage, three points of the function $g_3 \equiv \psi(\theta)$ are considered, taking into account the roots of Chebyshev's polynomial

(Chebyshev, 1955):

$$\alpha_1 = \frac{\alpha_{max} + \alpha_{min}}{2} - \frac{\alpha_{max} - \alpha_{min}}{2} \cos \frac{\pi}{6}$$

$$\alpha_2 = \frac{\alpha_{max} + \alpha_{min}}{2}$$

$$\alpha_3 = \frac{\alpha_{max} + \alpha_{min}}{2} - \frac{\alpha_{max} - \alpha_{min}}{2} \cos \frac{5\pi}{6}$$

$$\theta_1 = g_2(g_1(\alpha_1))$$

$$\theta_2 = g_2(g_1(\alpha_2))$$

$$\theta_3 = g_2(g_1(\alpha_3))$$

$$\psi_1 = \psi(\alpha_1) = k_2 \left[\ln \left(\frac{\alpha_1}{k_1} + 1 \right) - \ln 1 \right] - 2\pi/6$$

$$\psi_2 = \psi(\alpha_2) = k_2 \left[\ln \left(\frac{\alpha_2}{k_1} + 1 \right) - \ln 1 \right] - 2\pi/6$$

$$\psi_3 = \psi(\alpha_3) = k_2 \left[\ln \left(\frac{\alpha_3}{k_1} + 1 \right) - \ln 1 \right] - 2\pi/6$$

Considering the governing equation of the linkage

$$h(\theta, \psi^*, L_2, L_3, L_4) = 0$$

wherein L_2, L_3, L_4 are the lengths of the three bars (L_1 is considered equal to 1), a system of three equations and three unknowns may be stated (Waldron & Kinzel, 2004):

$$h_1(\theta_1, \psi_1^*, L_2, L_3, L_4) = 0$$

$$h_2(\theta_2, \psi_2^*, L_2, L_3, L_4) = 0$$

$$h_3(\theta_3, \psi_3^*, L_2, L_3, L_4) = 0$$

wherein $\psi_1^* = \psi_1$, $\psi_2^* = \psi_2$, $\psi_3^* = \psi_3$.

Solution of the system provides the lengths L_2, L_3, L_4 as

$$L_2 = 6.33822; \quad L_3 = 1.27937; \quad L_4 = 6.07355;$$

which may be scaled to any required dimension. For example, using a scale of 40,

$$L_1 = 40.00000; \quad L_2 = 253.52880; \quad L_3 = 51.17494; \quad L_4 = 242.94184;$$

Solution of the governing equation provides $\psi^*(\theta) \equiv g_3^*$. Functions g_3^* and g_3 coincide each other at points (θ_1, ψ_1), (θ_2, ψ_2), (θ_3, ψ_3).

The error function of the mechanism based on two pairs of noncircular gears and one four-bar linkage may be obtained as

$$\Delta \psi^* = \psi^*(\alpha) - \psi(\alpha) \tag{10.6.7}$$

If we consider now that function $\psi(\alpha)$ must be provided by just a single four-bar linkage and we notify the function generated by the linkage as $\psi^{**}(\alpha)$, the error function is obtained as

$$\Delta\psi^{**} = \psi^{**}(\alpha) - \psi(\alpha) \tag{10.6.8}$$

Figure 10.6.1(a) shows the two pairs of noncircular gears coupled with the linkage at position (θ_1, ψ_1). Figure 10.6.1(b) shows the error functions $\psi^*(\alpha) - \psi(\alpha)$ and $\psi^{**}(\alpha) - \psi(\alpha)$. The standard deviation of error function $\psi^*(\alpha) - \psi(\alpha)$ is 0.0545, whereas the standard deviation of error function $\psi^{**}(\alpha) - \psi(\alpha)$ is 0.1430. The standard deviation may vary depending on chosen points (α_i, ψ_i), $(i = 1, 2, 3)$; it depends as well on intervals of rotations for angles α and ψ.

10.6.2 Tandem Design of Slider-Crank Mechanism Coupled with Modified Elliptical Gears

10.6.2.1 Preliminary Information

Tandem design of the linkage means that the rotation of the crank is provided by a pair of two modified elliptical gears with centrodes σ_1 and σ_2 (Fig. 10.6.2(a)). Figure 10.6.2(b) shows two functions: (i) symmetric function $s(\alpha)$ of slider 5 of a conventional crank-slider linkage, and (ii) asymmetric function $s(\phi_1)$ of the linkage coupled with the pair of gears 1 and 2 (centrodes σ_1 and σ_2).

Here, s represents the instantaneous position of slider 5; α determines the instantaneous position of the crank; ϕ_1 is the current position of centrode 1 with respect to the center distance $O_1 - O_2$ of the centrodes.

Function $s(\phi_1)$ of slider 5 (Fig. 10.6.2(a)) corresponds to two cycles of the work of the linkage (Fig. 10.6.2(b)): the working cycle with lower velocity of the slider, and the free-running cycle with greater velocity. Such variation of the speed of the slider is obtained by application of modified elliptical gears (Fig. 10.6.2(a)).

Function $s(\alpha)$ of a conventional crank-slider linkage with symmetric cycles (Fig. 10.6.3) is determined as

$$s(\alpha) = -a\cos\alpha + \sqrt{b^2 - a^2\sin^2\alpha} \tag{10.6.9}$$

where a and b are the lengths of the crank and connecting rod, respectively.

10.6.2.2 Basic Ideas of Modification of Elliptical Centrodes

The modification of elliptical centrodes is illustrated with Fig. 10.6.4, which shows modification of upper and lower branches of the modified centrode (Figs. 10.6.4(a) and 10.6.4(b), respectively).

Position vector $\overline{O_1 A_2}$ of upper modified centrode (Fig. 10.6.4(a)) is determined as

$$|\overline{O_1 A_2}| = |\overline{O_1 A_1}|, \qquad \frac{\phi_1}{m_I} \neq \phi_1 \tag{10.6.10}$$

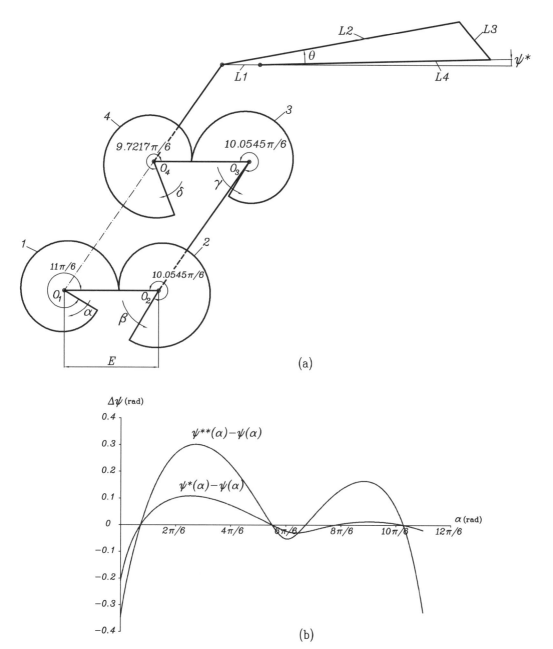

Figure 10.6.1. (a) Schematic illustration of four-bar linkage coupled with two pairs of non-circular gears for generation of function $z = \ln u$, $1 < u < 10$; (b) functions $\psi^*(\alpha) - \psi(\alpha)$ and $\psi^{**}(\alpha) - \psi(\alpha)$.

where m_I is the coefficient used for modification of polar angle ϕ_1. Similarly, Fig. 10.6.4(b) shows the position vector $\overline{O_1 B_2}$ of the lower branch of modified centrodes, determined as

$$|\overline{O_1 B_2}| = |\overline{O_1 B_1}|, \qquad \frac{2\pi - \phi_1}{m_{II}} \neq 2\pi - \phi_1 \qquad (10.6.11)$$

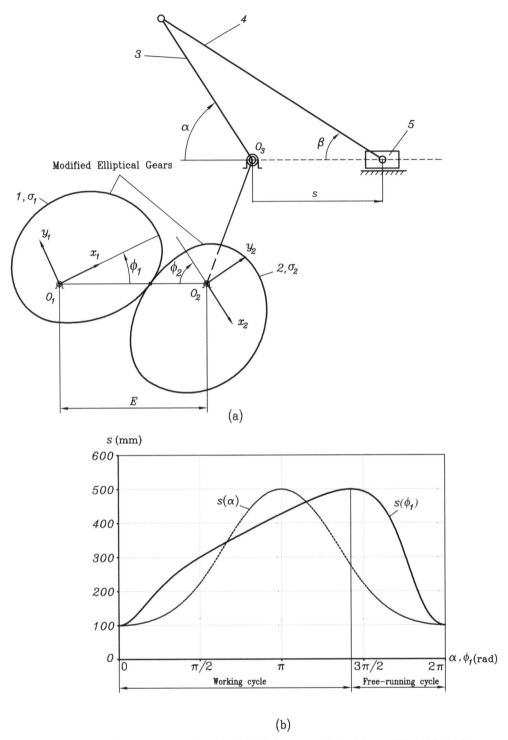

Figure 10.6.2. (a) Illustration of crank-slider linkage coupled with modified elliptical gears; (b) function $s(\alpha)$ for a conventional linkage and function $s(\phi_1)$ for the linkage coupled with modified elliptical gears.

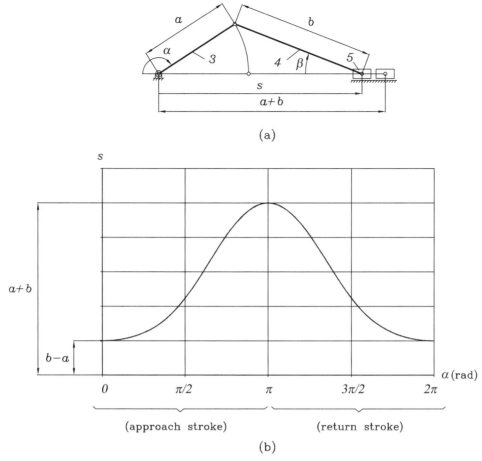

Figure 10.6.3. Crank-slider mechanism: (a) illustration of links 3, 4, and 5; (b) illustration of governing function $s(\alpha)$.

Coefficients m_{II} and m_I of modification of polar angles of the lower and upper branches are related as

$$m_{II} = \frac{m_I}{2m_I - 1} \qquad (10.6.12)$$

Branches $\sigma_1^{(I)}$ and $\sigma_1^{(II)}$ are shown in Fig. 10.6.4(c).

10.6.2.3 Analytical Determination of Modified Elliptical Centrodes

The design of the elliptical centrodes is based on the application of derivative functions (Fig. 10.6.5)

$$m_{21}^{(I)}(\phi_1) = \frac{1 - e^2}{1 + e^2 - 2e\cos(m_I\phi_1)}, \qquad 0 \le \phi_1 \le \frac{\pi}{m_I} \quad (10.6.13)$$

$$m_{21}^{(II)}(\phi_1) = \frac{1 - e^2}{1 + e^2 - 2e\cos(m_{II}(2\pi - \phi_1))}, \qquad \frac{\pi}{m_I} \le \phi_1 \le 2\pi \quad (10.6.14)$$

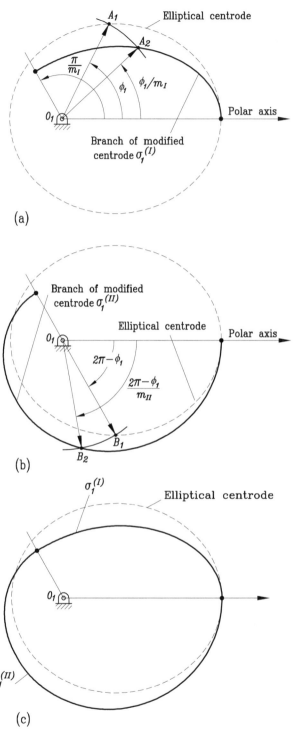

Figure 10.6.4. Illustration of conventional elliptical centrode and the two branches $\sigma_1^{(I)}$ and $\sigma_1^{(II)}$ of the modified centrode.

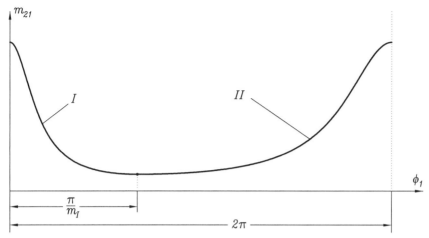

Figure 10.6.5. Illustration of asymmetric derivative functions $m_{21}^{(i)}(\phi_1)$, $i = I, II$.

The modified elliptical centrode σ_1 of gear 1 (Fig. 10.6.2(a)) is determined with

$$r_1(\phi_1) = \frac{p}{1 - e\cos(m_I\phi_1)}, \qquad 0 \le \phi_1 \le \frac{\pi}{m_I} \qquad (10.6.15)$$

$$r_1(\phi_1) = \frac{p}{1 - e\cos(m_{II}(2\pi - \phi_1))}, \qquad \frac{\pi}{m_I} \le \phi_1 \le 2\pi \qquad (10.6.16)$$

The centrode of gear 2 is determined with

$$r_2(\phi_1) = E - r_1(\phi_1) \qquad (10.6.17)$$

$$\phi_2(\phi_1) = \int_0^{\phi_1} m_{21}(\phi_1)d\phi_1 \qquad (10.6.18)$$

10.6.2.4 Numerical Examples

Two examples of tandem design are considered with function $s(\phi_1)$ shown in Fig. 10.6.2(b), and with function $s(\phi_1)$ shown in Fig. 10.6.6(b).

The dimensions of the links of the crank-slider linkages for both cases of design are $a = 200$ mm and $b = 300$ mm. For case of design 1 (Fig. 10.6.2), the parameters of design of the elliptical centrodes are: major axis of ellipse $2a = 120$ mm; eccentricity $e = 0.5$, coefficient of modification of branch $\sigma_1^{(I)}$, $m_I = 1.5$, and center distance $E = 2a = 120$ mm. For the case of design 2 (Fig. 10.6.6), link 3 of the crank-slider linkage is mounted with an offset angle α_0 with respect to the initial position of gear 2, where $\alpha_0 = 2\pi/3$.

Figure 10.6.2(a) shows the centrodes of the modified elliptical gears, and Fig. 10.6.2(b) shows functions $s(\alpha)$ and $s(\phi_1)$ of the slider-crank linkage. Longer approach and faster return strokes are obtained. The positions of the crank and centrode 2 are synchronized because $\alpha = \phi_2$ (Fig. 10.6.2(b)).

In case of design 2 (Fig. 10.6.6(a)), there is an offset angle α_0 formed by the initial positions of crank 3 and gear 2. Such assembly affects the shape of governing function $s(\phi_1)$, as shown in Fig. 10.6.6(b).

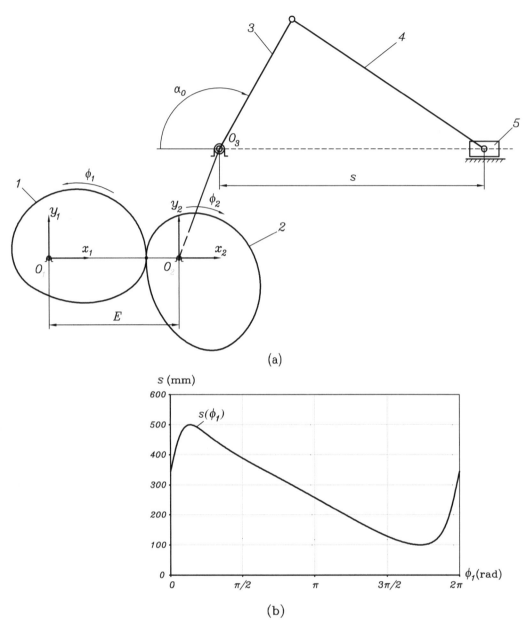

Figure 10.6.6. (a) Illustration of assembly of crank-slider mechanism with modified elliptical gears; (b) function $s(\phi_1)$.

10.6.3 Tandem Design of Scotch-Yoke Mechanism Coupled with Noncircular Gears

The Scotch-Yoke mechanism in the proposed design is driven by a pair of noncircular gears 1 and 2 and two circular gears 3 and 4 (Fig. 10.6.7(a)).

The purpose of tandem design of the Scotch-Yoke mechanism (coupled with gears) is to obtain an improved function of output velocity $v_7(\phi_1)$ (see Fig. 10.6.8) with the following conditions:

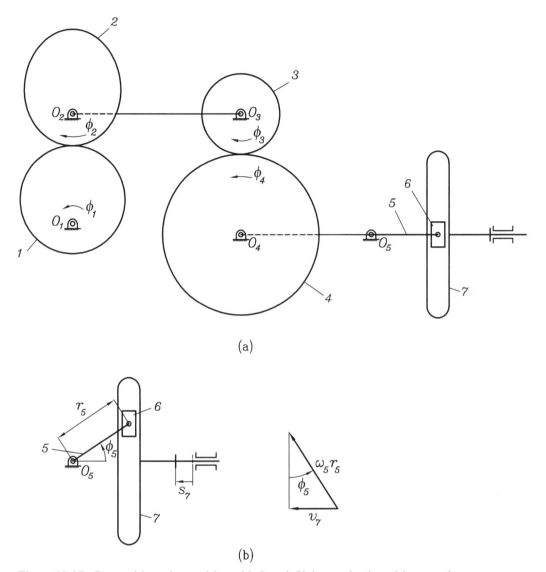

Figure 10.6.7. Composition of gear drive with Scotch-Yoke mechanism: (a) eccentric gear drive and Scotch-Yoke mechanism at their initial positions; (b) links of Scotch-Yoke mechanism and detail of related velocities.

(i) Gear 1 (Fig. 10.6.7(a)) is an eccentric involute gear and noncircular gear 2 is conjugated to gear 1.

(ii) Gears 3 and 4 (Fig. 10.6.7(a)) are circular gears with gear ratio of 2 $(\omega^{(3)}/\omega^{(4)}) = 2$); gear 4 performs one revolution for two revolutions of gear 3.

(iii) Function $v_7(\phi_1)$ is of a period $\phi_1 = 2\pi$, and this is obtained by application of circular gears 3 and 4 in addition to eccentric circular gear 1 and noncircular gear 2.

The basic kinematic relation of the coupled mechanism is based on the drawings of Fig. 10.6.7. Figure 10.6.7(a) shows the initial position of links at $\phi_1 = 0$.

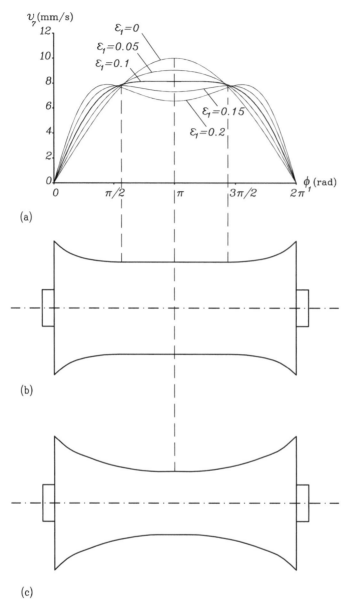

(a)

(b)

(c)

Figure 10.6.8. Results from application of tandem design based on an eccentric gear drive and Scotch-Yoke mechanism: (a) velocity of the Scotch-Yoke as a function of angle of rotation of the eccentric gear for different values of the parameter of eccentricity ε_1; (b) textile ball when $\varepsilon_1 = 0.1$ and the shape of the textile ball is improved; and (c) textile ball for a Scotch-Yoke mechanism without noncircular gears.

The basic kinematic relations of the velocities of the links of the mechanism are represented by

$$v_7 = \omega_5 r_5 \sin \phi_5 \qquad (10.6.19)$$

wherein ϕ_5 is the angle of rotation of the crank. Such a relation is a sine-type function.

Angular velocity ω_5 is determined as

$$\omega_5 = \omega_4 = m_{43}\omega_3 = m_{43}\omega_2 = m_{43}m_{21}(\phi_1)\omega_1 \tag{10.6.20}$$

wherein $m_{21}(\phi_1)$ is determined as (see Litvin *et al.*, 2008)

$$m_{21} = \frac{c_1}{c_1 - \left(1 - \varepsilon_1^2 \sin^2 \phi_1\right)^{\frac{1}{2}} - \varepsilon_1 \cos \phi_1} - 1 \tag{10.6.21}$$

Here, $c_1 = \dfrac{E}{r_{p1}}$, $\varepsilon_1 = \dfrac{e_1}{r_{p1}}$, wherein E is the center distance of the eccentric gear drive, r_{p1} is the pitch radius of the eccentric gear, e_1 is the eccentricity (see Litvin *et al.*, 2008).

Angle of rotation ϕ_5 is also derived as

$$\phi_5 = \phi_4 = m_{43}\phi_3 = m_{43}\phi_2 = m_{43}\left(\int_0^{2\pi} m_{21}d\phi_1\right) \tag{10.6.22}$$

Equations (10.6.20), (10.6.21), and (10.6.22) yield

$$v_7 = m_{43}m_{21}\omega_1 r_5 \sin\left(m_{43}\left(\int_0^{2\pi} m_{21}d\phi_1\right)\right) \tag{10.6.23}$$

Figure 10.6.8(a) shows the velocity function $v_7(\phi_1)$ for various values of parameter of eccentricity ε_1. A flat function is obtained with $\varepsilon_1 \approx 0.1$. Figures 10.6.8(b) and (c) show textile balls when $\varepsilon_1 = 0.1$ and $\varepsilon_1 = 0$. Figure 10.6.8(b) shows an optimized solution for the volume of wrapped strap around the ball with respect to the solution shown in Fig. 10.6.8(c).

Application of elliptical gears instead of an eccentric gear drive is also possible. Tandem design based on two pairs of elliptical gears and a Scotch-Yoke mechanism provides the following expression for the velocity v_7:

$$v_7 = m_{43}m_{21}\omega_1 r_5 \sin\left(m_{43} \cdot \arctan\left(\frac{1+e}{1-e} \tan \frac{\phi_1}{2}\right)\right) \tag{10.6.24}$$

wherein m_{21} is given by (see Litvin *et al.*, 2007)

$$m_{21} = \frac{1 - e^2}{1 + e^2 - 2e \cos \phi_1} \tag{10.6.25}$$

and e is the eccentricity of the ellipse.

Figure 10.6.9 shows the velocity function for different values of the parameter of eccentricity e_1 when a tandem design based on one pair of elliptical gears and a Scotch-Yoke mechanism is applied.

10.6.4 Tandem Design of Mechanism Formed by Two Pairs of Noncircular Gears and Racks

The assembly of the gears and racks are shown in Figs. 10.6.10 and 10.6.11. Such mechanisms allow us to increase the magnitude of translational motion.

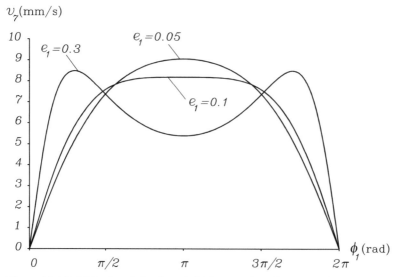

Figure 10.6.9. Velocity of the Scotch-Yoke as a function of angle of rotation of an elliptical gear for different values of the parameter of eccentricity e_1.

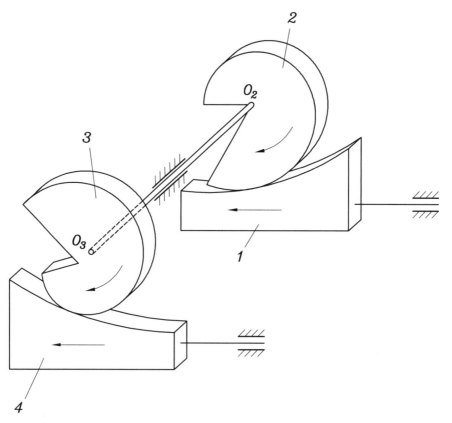

Figure 10.6.10. Two pairs of gears with racks 1 and 4 and noncircular gears 2 and 3.

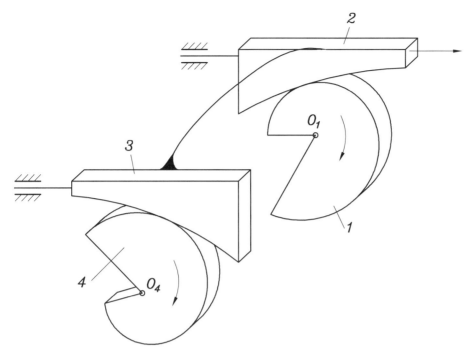

Figure 10.6.11. Two pairs of gears with noncircular gears 1 and 4 and racks 2 and 3.

The basic kinematic relations are obtained as follows. Figure 10.6.12(a) shows that rotation of link 1 with angular velocity $\omega^{(1)}$ is transformed into translation of link 2 with linear velocity $v^{(2)}$. Point I of tangency of centrodes 1 and 2 is located on line $O_1 n$ that is perpendicular to $x^{(2)}$; point I moves along $O_1 n$ in the process of transformation of motions (Figs. 10.6.12(a) and (c)). Point I traces in coordinate systems rigidly connected to links 1 and 2 centrodes σ_1 and σ_2 of the drive (Figs. 10.6.12(b) and (c)).

Centrode σ_1 is a polar curve (Fig. 10.6.12(b)), $O_1 A_1$ is the polar axis, and $\mathbf{r}_1(\phi_1)$ is the position vector of σ_1.

Centrode σ_2 is represented in coordinate system $S_2(x_2, y_2)$, and the position vector $\mathbf{r}_2(\phi_1)$ of σ_2 is represented by two components: $\overline{O_2 O_1} = \int_0^{\phi_1} ds_2$, and $\overline{O_1 I}$ (Fig. 10.6.12(c)) Centrodes σ_1 and σ_2 roll over each other in the process of meshing.

Centrode σ_1 of noncircular gear is a polar curve, and the position vector $r_1(\phi_1) = \overline{O_1 I}$ forms angle ϕ_1 with the polar axis $\overline{O_1 A_1}$. The magnitude of $r_1(\phi_1)$ is determined by the ratio

$$r_1(\phi_1) = \frac{v^{(2)}}{\omega^{(1)}} = \frac{ds_2}{d\phi_1} \qquad (10.6.26)$$

The position vector $\overline{O_2 I}$ of centrode σ_2 is determined as

$$r_2(\phi_1) = \int_0^{\phi_1} ds_2 \mathbf{i}_2 + r_1(\phi_1)\mathbf{j} \qquad (10.6.27)$$

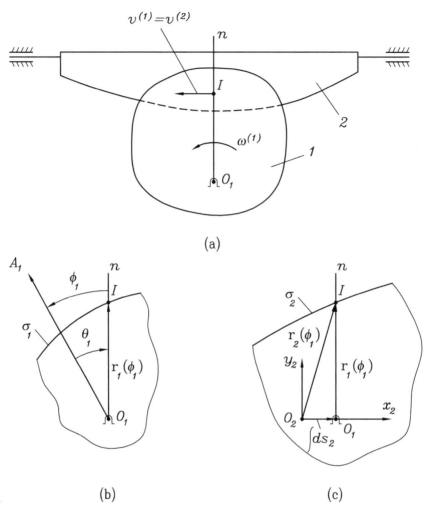

$$v^{(1)} = v^{(2)}$$

(a)

(b)   (c)

Figure 10.6.12. Toward derivation meshing of noncircular gear and rack: (a) illustration of transformation of motions; (b) centrode 1 and polar axis A_1; (c) centrode 2 and coordinate system (x_2, y_2).

10.6.4.1 Generation of Function

A gear drive formed by a rack and noncircular gear may be applied for generation of function $z = z(u)$, $u_{min} < u < u_{max}$. The angle of rotation of centrode σ_1 and displacement of centrode σ_2 are represented as

$$\phi_1 = k_1(u - u_{min}) + \phi_{1,min}, \quad s_2 = k_2[z(u) - z(u_{min})] + s_{2,min} \qquad (10.6.28)$$

Centrode σ_1 (of noncircular gear) is a polar curve ($\overline{O_1 A_1}$ is the polar axis) and is represented (Fig. 10.6.12) as

$$r_1(u) = \frac{k_2}{k_1}[z'(u)], \quad \phi_1(u) = k_1(u - u_{min}) + \phi_{1,min} \qquad (10.6.29)$$

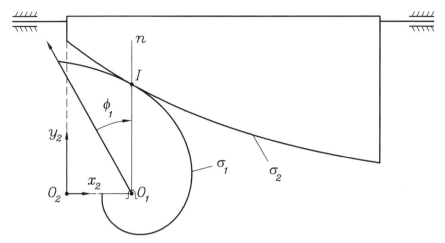

Figure 10.6.13. For generation of function $z = \ln u$, $1 < u < 5$, by a gear drive formed by a rack and noncircular gear.

Centrode σ_2 (of the rack) is represented in coordinate system (y_2, x_2) as

$$x_2(u) = k_2[z(u) - z(u_{min})] + s_{2,min}, \quad y_2(u) = r_1(u) - r_1(u_{min}) = \frac{k_2}{k_1}[z'(u) - z'(u_{min})]$$

(10.6.30)

Coefficients k_1 and k_2 are the scale ones. Centrode σ_1 is usually an unclosed curve, and the maximal angle of rotation of the noncircular gear is $\phi_{1,max} < 2\pi$.

NUMERICAL EXAMPLE 10.6.1: GENERATION OF FUNCTION $z = \ln u$, $1 < u < 5$, BY A GEAR DRIVE FORMED BY A RACK AND NONCIRCULAR GEAR. The procedure of determination of centrodes is as follows:

(1) Intervals of rotation for centrode σ_1 and displacement for centrode σ_2 are

$$0 < \phi_1 < \frac{10\pi}{6}$$ (10.6.31)

$$0 < s_2 < 50\frac{10\pi}{6}$$ (10.6.32)

(2) Coefficients k_1 and k_2 are determined as

$$k_1 = \frac{\phi_{1,max} - \phi_{1,min}}{u_{max} - u_{min}} = \frac{10\pi/6}{5-1}$$ (10.6.33)

$$k_2 = \frac{s_{2,max} - s_{2,min}}{\ln u_{max} - \ln u_{min}} = \frac{50\frac{10\pi}{6}}{\ln 5}$$ (10.6.34)

(3) Applications of Eqs. (10.6.29) and (10.6.30) provide representation of centrodes σ_1 and σ_2, respectively (see Fig. 10.6.13).

11 Additional Numerical Problems

Problem 11.1. Function $f(x) = x$, $1 \le x \le 10$, to be generated by a mechanism of noncircular gears mounted with a center distance of $E = 100$ mm is given. Compute the centrodes and represent their graphs. The centrodes will be non-closed curves wherein $(\phi_1)_{max} = (\phi_2)_{max} = 10\pi/6$.

According to Eqs. (2.4.6) and (2.4.7), the angles of rotation of the driving and driven gears are

$$\phi_1 = k_1(x - 1) \tag{11.1.1}$$

$$\phi_2 = k_2[f(x) - f(1)] = k_2[x - 1] \tag{11.1.2}$$

wherein $f(x) = x$ and $f(1) = 1$, according to the given function. The gear ratio is determined as (Eq. (2.4.8))

$$m_{12}(x) = \frac{d\phi_1}{d\phi_2} = \frac{k_1}{k_2 f'(x)} = \frac{k_1}{k_2} \tag{11.1.3}$$

As mentioned previously, $(\phi_1)_{max} = (\phi_2)_{max} = \dfrac{10\pi}{6}$ for $x = 10$, so

$$(\phi_1)_{max} = \frac{10\pi}{6} = k_1(10 - 1) = 9k_1 \tag{11.1.4}$$

and therefore

$$k_1 = \frac{10\pi}{54} \tag{11.1.5}$$

Similarly, constant k_2 is obtained as

$$k_2 = \frac{10\pi}{54} \tag{11.1.6}$$

The gear ratio according to Eq. (11.1.3) is

$$m_{12}(x) = \frac{d\phi_1}{d\phi_2} = \frac{k_1}{k_2} = 1 \tag{11.1.7}$$

as expected for generation of the lineal function $f(x) = x$.

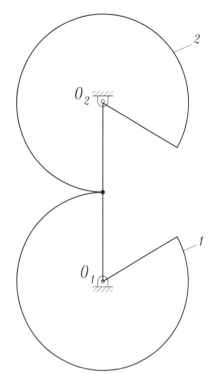

Figure 11.1.1. Centrodes of gears generating function $f(x) = x$.

Centrode 1 is determined by

$$r_1(\phi_1) = \frac{E}{1 + m_{12}(\phi_1)} = \frac{E}{2} \tag{11.1.8}$$

which corresponds to a circle of constant radius $E/2$. Centrode 2 is determined by

$$r_2(\phi_1) = E \frac{m_{12}(\phi_1)}{1 + m_{12}(\phi_1)} = E\frac{1}{2} = \frac{E}{2} \tag{11.1.9}$$

which also corresponds to a circle of constant radius $E/2$. The rotation angles of gear 2 and 1 are determined by

$$\phi_2(\phi_1) = \int_0^{\phi_1} \frac{d\phi_1}{1} = \int_0^{\phi_1} d\phi_1 = \phi_1 \tag{11.1.10}$$

The centrodes corresponding to generation of function $f(x) = x$ are represented by Fig. 11.1.1.

Problem 11.2. Function $f(x) = x^2$, $1 \le x \le 10$, to be generated by a mechanism of noncircular gears is given. Compute the centrodes and represent their graphs for the same conditions that for Problem 11.1, wherein $E = 100$ mm and $(\phi_1)_{max} = (\phi_2)_{max} = 10\pi/6$.

The angles of rotation of the driving and driven gears are

$$\phi_1 = k_1(x - 1) \qquad (11.2.1)$$

$$\phi_2 = k_2[x^2 - 1] \qquad (11.2.2)$$

wherein $f(x) = x^2$ and $f(x_1) = 1$.

According to the conditions of the problem, $(\phi_1)_{max} = (\phi_2)_{max} = \dfrac{10\pi}{6}$ for $x = 10$, so

$$(\phi_1)_{max} = \frac{10\pi}{6} = k_1(10 - 1) = 9k_1 \qquad (11.2.3)$$

and therefore

$$k_1 = \frac{10\pi}{6 \cdot 9} \qquad (11.2.4)$$

Similarly,

$$(\phi_2)_{max} = \frac{10\pi}{6} = k_2(10^2 - 1) = k_1 \cdot 99 \qquad (11.2.5)$$

and therefore

$$k_2 = \frac{10\pi}{6 \cdot 99} \qquad (11.2.6)$$

The gear ratio is given by

$$m_{12}(x) = \frac{d\phi_1}{d\phi_2} = \frac{k_1}{k_2 f'(x)} = \frac{k_1}{2k_2 x} \qquad (11.2.7)$$

From Eq. (11.2.1)

$$x = \frac{\phi_1}{k_1} + 1 \qquad (11.2.8)$$

so that the gear ratio may be expressed by a function of ϕ_1 as

$$m_{12}(\phi) = \frac{d\phi_1}{d\phi_2} = \frac{k_1}{2k_2 x} = \frac{k_1^2}{2k_1 k_2 + 2k_2 \phi_1} = \frac{k_1^2}{2k_2(k_1 + \phi_1)} \qquad (11.2.9)$$

Centrode 1 is determined as

$$r_1(\phi_1) = \frac{E}{1 + m_{12}(\phi_1)} \qquad (11.2.10)$$

where $m_{12}(\phi_1)$ is determined by Eq. (11.2.9). Centrode 2 is determined as

$$r_2(\phi_1) = E \frac{m_{12}(\phi_1)}{1 + m_{12}(\phi_1)} \qquad (11.2.11)$$

The rotation angles of gear 2 and 1 are related by

$$\phi_2(\phi_1) = \int_0^{\phi_1} \frac{d\phi_1}{m_{12}(\phi_1)} = \int_0^{\phi_1} \frac{2k_2 x}{k_1} d\phi_1 = \int_0^{\phi_1} \frac{2k_2 \left(\frac{\phi_1}{k_1} + 1\right)}{k_1} d\phi_1 = \frac{k_2 \phi_1(\phi_1 + 2k_1)}{k_1^2}$$

$$(11.2.12)$$

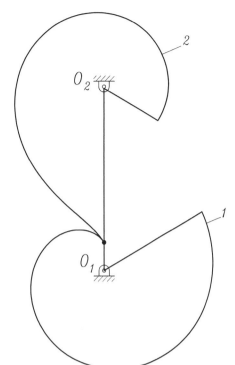

Figure 11.2.1. Centrodes of gears generating function $f(x) = x^2$.

The centrodes corresponding to generation of function $f(x) = x^2$ are represented in Fig. 11.2.1.

Problem 11.3. Derive the centrodes for generation of function

$$\tan\left(\frac{\phi_2}{2}\right) = \left(\frac{1-e}{1+e}\right)\tan\left(\frac{\phi_1}{2}\right)$$

Consider $\phi_{1\,max} = \phi_{2\,max} = 2\pi$ and $E = 100$ mm.

According to Section 2.4, we meet the conditions of Case 1 wherein the derivative function $m_{12}(\phi_1) = d\phi_1/d\phi_2$ and the center distance of the gear drive with noncircular gears are given or known. The to-be-generated function is transformed into

$$\phi_2 = 2\,\arctan\left[\left(\frac{1-e}{1+e}\right)\tan\left(\frac{\phi_1}{2}\right)\right] \tag{11.3.1}$$

allowing further derivations.

The derivative $d\phi_2/d\phi_1$ must be determined toward determination of the derivative function $m_{12}(\phi_1)$

$$\frac{d\phi_2}{d\phi_1} = \frac{1-e^2}{1+e^2+2\,e\,\cos\phi_1} \tag{11.3.2}$$

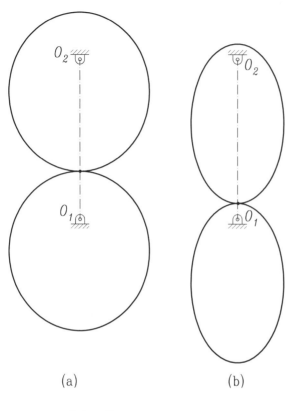

Figure 11.3.1. Centrodes of noncircular gears generating the function $\tan\left(\frac{\phi_2}{2}\right) = \left(\frac{1-e}{1+e}\right)\tan\left(\frac{\phi_1}{2}\right)$ for two cases of design: (a) $e = 0.4$ and (b) $e = 0.8$, wherein $E = 100$ mm.

(a)   (b)

and therefore

$$m_{12}(\phi_1) = \frac{d\phi_1}{d\phi_2} = \frac{1 + e^2 + 2\,e\,\cos\phi_1}{1 - e^2} \tag{11.3.3}$$

Centrode σ_1 is determined by

$$r_1(\phi_1) = \frac{E}{1 + m_{12}(\phi_1)} \tag{11.3.4}$$

Centrode σ_2 is determined by

$$r_2(\phi_1) = E\frac{m_{12}(\phi_1)}{1 + m_{12}(\phi_1)} \tag{11.3.5}$$

$$\phi_2(\phi_1) = \int_0^{\phi_1} \frac{d\phi_1}{m_{12}(\phi_1)} = 2\,\arctan\left[\left(\frac{1-e}{1+e}\right)\tan\left(\frac{\phi_1}{2}\right)\right] \tag{11.3.6}$$

Figure 11.3.1 shows the centrodes of noncircular gears generating the function $\tan\left(\frac{\phi_2}{2}\right) = \left(\frac{1-e}{1+e}\right)\tan\left(\frac{\phi_1}{2}\right)$ for two cases of design wherein the eccentricity of the ellipse is (a) $e = 0.4$ and (b) $e = 0.8$. For all cases of design, a center distance $E = 100$ mm has been considered.

Problem 11.4. Derive the centrodes for generation of the function

$$\phi_2 = b\phi_1 - a\cos\phi_1$$

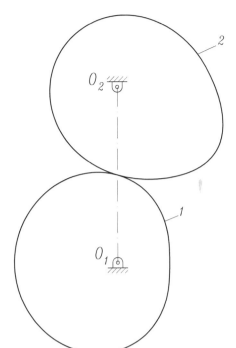

Figure 11.4.1. Centrodes of noncircular gears generating the function $\phi_2 = b\phi_1 - a\cos\phi_1$ wherein $b/a = 1.8$, $a = 1/1.8$, and $E = 100$ mm.

Choose $b/a = 1.8$ and determine a for providing a closed centrode for gear 2.

Solution:

$$\frac{d\phi_2}{d\phi_1} = b + a\sin\phi_1 \tag{11.4.1}$$

$$m_{12}(\phi_1) = \frac{d\phi_1}{d\phi_2} = \frac{1}{b + a\sin\phi_1} \tag{11.4.2}$$

$$r_1(\phi_1) = \frac{E}{1 + m_{12}(\phi_1)}, \qquad r_2(\phi_1) = E\frac{m_{12}(\phi_1)}{1 + m_{12}(\phi_1)} \tag{11.4.3}$$

$$\phi_2(\phi_1) = \int_0^{\phi_1} \frac{d\phi_1}{m_{12}(\phi_1)} = a + b\phi_1 - a\cos\phi_1 \tag{11.4.4}$$

Equation (11.4.4), considering the conditions of design of centrodes as closed form curves (Section 2.6), yields

$$\phi_2 = 2\pi = \int_0^{2\pi} \frac{d\phi_1}{m_{12}(\phi_1)} = a + 2\pi b - a = 2\pi b \tag{11.4.5}$$

Therefore, to obtain a closed form centrode, $b = 1$, and according to the problem condition $b/a = 1.8$, we obtain $a = 1/1.8$. The centrodes of gear 1 and gear 2 are represented in Fig. 11.4.1.

Problem 11.5. Following functions are given:

(a) $\dfrac{d\phi_2}{d\phi_1} = a + b\cos\phi_1$

(b) $\dfrac{d\phi_2}{d\phi_1} = \dfrac{1}{a + b\cos\phi_1}$

Determine the centrodes of the respective noncircular gears that will generate such functions, and obtain the relation between parameters a and b to obtain closed centrodes for gear 1 and 2. Consider the center distance $E = 100$ mm.

For function (a), the gear ratio is given by

$$m_{12}(\phi_1) = \frac{d\phi_1}{d\phi_2} = \frac{1}{a + b\cos\phi_1} \qquad (11.5.1)$$

The relation between the angles of rotation of gear 1 and gear 2 is given by

$$\phi_2(\phi_1) = \int_0^{\phi_1} \frac{d\phi_1}{m_{12}(\phi_1)} = a\,\phi_1 + b\,\sin\phi_1 \qquad (11.5.2)$$

Equation (11.5.2), for $\phi_1 = 2\pi$, $\phi_2 = 2\pi$, is the condition of obtaining a closed centrode for gear 2. Equation (11.5.2) with $\phi_1 = \phi_2 = 2\pi$ yields $a = 1.0$. The shape of the centrodes depends on parameter b. Figure 11.5.1 shows centrodes for different values of parameter b.

We consider as well generation of $\phi_2 = a\phi_1 - b\sin\phi_1$. The difference with respect to generation of $\phi_2 = a\phi_1 + b\sin\phi_1$ is just the position of the polar axis that is turned an angle of $180°$ (see Fig. 11.5.2).

For function (b), the gear ratio is given by

$$m_{12}(\phi_1) = \frac{d\phi_1}{d\phi_2} = a + b\cos\phi_1 \qquad (11.5.3)$$

The relation between the angles of rotation of gear 1 and gear 2 is given by

$$\phi_2(\phi_1) = \int_0^{\phi_1} \frac{d\phi_1}{m_{12}(\phi_1)} = \int_0^{\phi_1} \frac{d\phi_1}{a + b\cos\phi_1} \qquad (11.5.4)$$

The solution of the integral in Eq. (11.5.4), (see Dwight, 1961) is

$$\phi_2(\phi_1) = \int \frac{d\phi_1}{a + b\cos\phi_1} = \frac{2}{\sqrt{a^2 - b^2}}\arctan\frac{(a-b)\tan\frac{\phi_1}{2}}{\sqrt{(a^2 - b^2)}}, \qquad a^2 > b^2 \quad (11.5.5)$$

The conditions to obtain a closed centrode for gear 2 are $a^2 > b^2$ and $a^2 - b^2 = 1.0$.

The function represented in Problem 11.3 is a particular case of function (b) wherein parameter $a = (1 + e^2)/(1 - e^2)$ and parameter $b = 2e/(1 - e^2)$. Figure 11.5.3 shows the centrodes generating function (b) for different values of parameter b, observing for all cases that $a = \sqrt{1 + b^2}$.

Problem 11.6. For the same conditions of Problem 11.2, determine the curvatures of centrodes 1 and 2 and determine the condition of convexity of centrode 2. In case of obtaining a nonconvex centrode, determine the conditions of avoiding it.

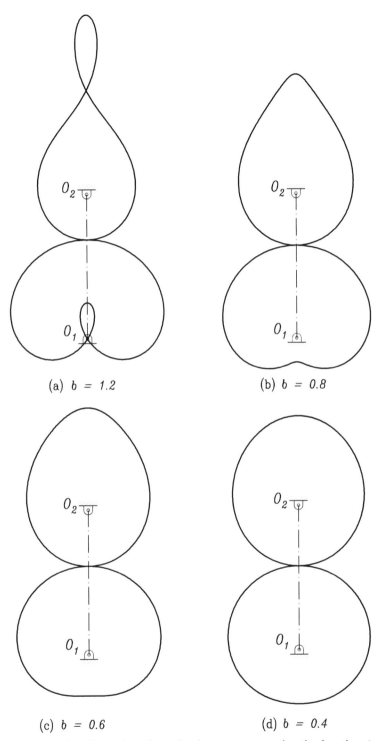

Figure 11.5.1. Centrodes of noncircular gears generating the function $\phi_2 = a\phi_1 + b\sin\phi_1$ for different values of parameter b, wherein $a = 1.0$ and center distance is $E = 100$ mm.

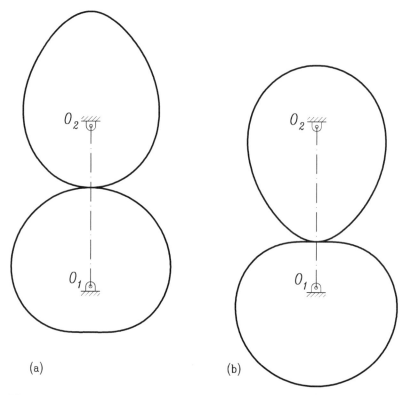

Figure 11.5.2. Centrodes of noncircular gears generating the function (a) $\phi_2 = a\phi_1 + b\sin\phi_1$ and (b) $\phi_2 = a\phi_1 - b\sin\phi_1$, wherein $a = 1.0$ and $b = 0.6$, and center distance is $E = 100$ mm.

The centrodes of the noncircular gears generating the function $f(x) = x^2$ are represented in Fig. 11.6.1(a). It may be observed that a small part of centrode 2 does not observe the condition of convexity.

The condition of convexity of centrodes is determined by observation of following conditions:

(i) For centrode 1 as

$$h_1(\phi_1) = 1 + m_{12}(\phi_1) + m_{12}''(\phi_1) \geq 0 \qquad (11.6.1)$$

(ii) For centrode 2 as

$$h_2(\phi_1) = 1 + m_{12}(\phi_1) + [m_{12}'(\phi_1)]^2 - m_{12}(\phi_1)m_{12}''(\phi_1) \geq 0 \qquad (11.6.2)$$

The conditions of convexity of centrode 2 as a function of ϕ_2 are represented in Fig. 11.6.1(b). The conditions of convexity for centrode 2 are not observed from $\phi_2 = 0$ to $\phi_2 = 0.217$ rad. Centrode 2 is concave for the mentioned interval of ϕ_2.

The curvatures of centrode 1 and 2 are represented in Figs. 11.6.1(c) and (d) as functions of ϕ_1 and ϕ_2, respectively. Figure 11.6.1(d) shows the following:

(i) The curvature changes its sign, and therefore centrode 2 has a point with curvature equal to zero.
(ii) The centrode at such a point changes its shape from concave to convex.

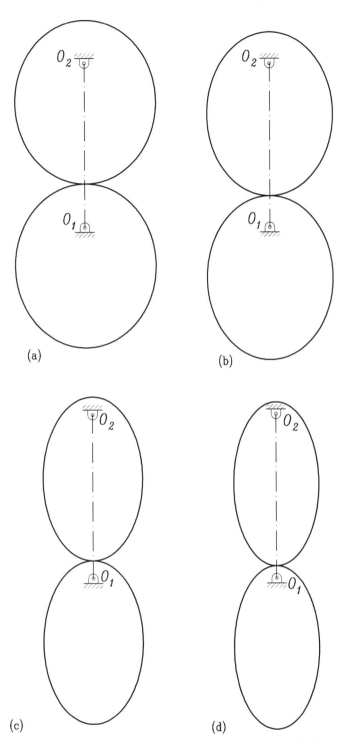

Figure 11.5.3. Centrodes of noncircular gears generating the function $\phi_2 = 1/(a\phi_1 + b\sin\phi_1)$ for different values of parameter b and center distance $E = 100$ mm, wherein parameter $a = \sqrt{1 + b^2}$.

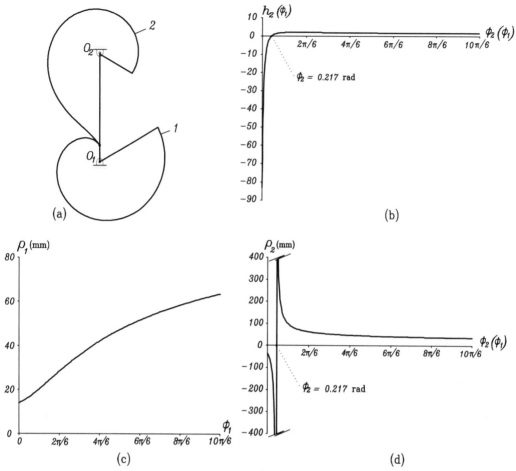

Figure 11.6.1. (a) Centrodes of noncircular gears generating function $z = u^2$; (b) representation of condition of concavity of centrode 2 according to Eq. (11.6.2); (c) curvature of centrode 1 as a function of ϕ_1; (d) curvature of centrode 2 as a function of ϕ_2.

Figure 11.6.1(b) shows that the condition of convexity equal to zero is obtained for $\phi_2 = 0.217$ rad. This condition coincides as well with the point wherein the curvature of centrode 2 changes its sign, represented by Fig. 11.6.1(b).

To avoid the area of concavity of centrode 2, three approaches may be applied:

(a) Change the interval of definition of function $z = u^2$. Currently, the interval $1 \leq u \leq 10$ is applied. By application of theory in Sections 2.8 and 2.9, it is observed that for the interval $1 \leq u \leq 4.62$, both centrodes 1 and 2 observe the conditions of convexity (Fig. 11.6.2(a))

(b) By extension of maximum angles of rotation of centrodes 1 and 2. For $\phi_{1max} = 25\pi/4$ and $\phi_{2max} = 25\pi/4$, centrode 2 observes the condition of convexity (Fig. 11.6.2(b))

(c) Using more than one pair of noncircular gears (see Section 10.5).

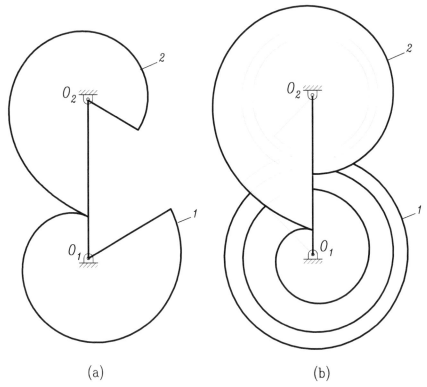

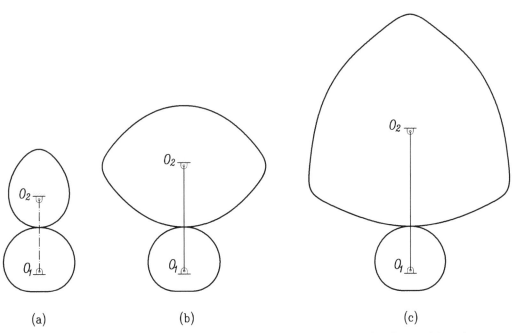

Figure 11.6.2. (a) Centrodes of noncircular gears generating function $z = u^2$ and observing the condition of convexity for both centrodes: (a) $0 \leq u \leq 4.62$, $\phi_{1 max} = 10\pi/6$, and $\phi_{2 max} = 10\pi/6$; (b) $0 \leq u \leq 10$, $\phi_{1 max} = 25\pi/4$, and $\phi_{2 max} = 25\pi/4$.

           (a)   (b)   (c)

Figure 11.7.1. Centrodes generating the function $\phi_2 = a\phi_1 + b\sin\phi_1$, wherein $a = 1.0$ and $b = 0.6$, for the following cases of design: (a) $n = 1$, (b) $n = 2$, (c) and $n = 3$ being n the number of revolutions of centrode 1 for one revolution of centrode 2.

Problem 11.7. Determine the centrodes generating the function $\phi_2 = a\phi_1 + b\sin\phi_1$, wherein $a = 1.0$ and $b = 0.6$ (see Fig. 11.5.2), for the following cases of design: (a) $n = 1$, (b) $n = 2$, and (c) $n = 3$ being n the number of revolutions of centrode 1 for 1 revolution of centrode 2.

The centrodes to-be-determined are represented in Fig. 11.7.1.

References

ARTOBOLEVSKI, IVAN IVANOVICH, LEVITSKII, N. I., & CERCUDINOV, S. A. 1959. *Synthesis of Planar Mechanisms*. Moscow: Fizmatghiz.

BOPP & REUTHER G.M.B.H. 1938. Improvements in or relating to milling toothed gears. *Patent No. DE668897, also published as GB531563*, December.

BOYD, W. W. 1940. Elliptical gears provide feed control. *Machine Design*, **12**.

BURMESTER, LUDWIG. 1888. *Lehrbuch der Kinematik*. Leipzig: Arthur Felix Verlag.

CHEBYSHEV, P. L. 1955. The Theory of Mechanisms Known as Parallelograms. *Selected Works, Publishing House of the USSR Academy of Sciences, Moscow*, 611–648.

COLLATZ, L. 1951. *Numerische Behandlung von Differentialgleichungen*. Berlin, Göttingen, Heidelberg: Springer.

DOONER, DAVID B. 2001. Function generation utilizing an eight-link mechanism and optimized non-circular gear elements with application to automotive steering. *Proceedings of the Institution of Mechanical Engineers, Part C: Journal of Mechanical Engineering Science*, **215**(7), 847–857.

DWIGHT, HERBERT B. 1961. *Tables of Integrals and Other Mathematical Data*. Prentice Hall; Fourth edition.

FELLOWS, EDWIN R. 1924 (November). *Gear generating cutting machine*. United States Patent 1516524.

FICHTENGOLZ, G. M. 1958. *Course in Differential and Integral Calculus*. 4 edn. Vol. 1. Moscow-Leningrad: State House of Physics, Mathematic Literature (in Russian).

FREUDENSTEIN, F. 1955. Approximate Synthesis of Four-bar Linkages. *Transactions of ASME*, **77**, 853–861.

GOLBER, HYMAN E. 1939. Rollcurve Gears. *Transactions of ASME*, **61**, 223–231.

HARTENBERG, R. S., & DENAVIT, J. 1964. *Kinematic Synthesis of Linkages*. McGraw-Hill Inc.

KISLITZIN, S. G. 1955. Aproximate Solution of Equation $\varphi(\varphi(x)) = f(x)$. *Proceedings of Leningrad Pedagogical Institute*, **90**.

KORN, GRANINO A., & KORN, THERESA M. 1968. *Mathematics Handbook for Scientist and Engineers*. 2nd edn. McGraw-Hill, Inc.

LITVIN, FAYDOR L. 1956. *Noncircular Gears: Design, Theory of Gearing, and Manufacture*. 2nd edn. Gos Tech Isdat, Leningrad, Moscow (in Russian).

LITVIN, FAYDOR L., & FUENTES, ALFONSO. 2004. *Gear Geometry and Applied Theory*. Second edn. New York (USA): Cambridge University Press.

LITVIN, FAYDOR L., & KETOV, C. F. 1949. Planetary reducer. *USSR Certificate of Inventions*.

LITVIN, FAYDOR L., *et al.* 1949 to 1951. Generation of Noncircular Gears by Enveloping Process. *USSR Certificate of Inventions*.

LITVIN, FAYDOR L., HSIAO, C. L., WANG, J. C., & ZHOU, X. 1994. Computerized Simulation of Generation of Internal Involute Gears and Their Assembly. *Journal of Mechanical Design, Transactions of the ASME*, **116**(3), 683–689.

LITVIN, FAYDOR L., GONZALEZ-PEREZ, IGNACIO, YUKISHIMA, KENJI, FUENTES, ALFONSO, & HAYASAKA, KENICHI. 2007. Generation of planar and helical elliptical gears by application of rack-cutter, hob, and shaper. *Computer Methods in Applied Mechanics and Engineering*, **196**, 4321–4336.

LITVIN, FAYDOR L., GONZALEZ-PEREZ, IGNACIO, FUENTES, ALFONSO, & HAYASAKA, KENICHI. 2008. Design and investigation of gear drives with non-circular gears applied for speed variation and generation of functions. *Computer Methods in Applied Mechanics and Engineering*, **197**, 3783–3802.

LITVIN, FAYDOR L., GONZALEZ-PEREZ, IGNACIO, FUENTES, ALFONSO, & HAYASAKA, KENICHI. 2009. Tandem design of mechanisms for function generation and output speed variation. *Computer Methods in Applied Mechanics and Engineering*, **198**(5-8), 860–876.

MODLER, K. H., LOVASZ, E. C., BÄR, GERT, NEUMANN, R., PERJU, D., PERNER, M., & MARGINEANU, D. 2009. General method for the synthesis of geared linkages with non-circular gears. *Mechanism and Machine Theory*, **44,** 726–738.

OTTAVIANO, ERIKA, MUNDO, DOMENICO, DANIELI, GUIDO A., & CECCARELLI, MARCO. 2008. Numerical and experimental analysis of non-circular gears and cam-follower systems as function generators. *Mechanism and Machine Theory*, **43**(8), 996–1008.

SMITH, W. C. 1995. The Math of Noncircular Gearing. *Gear Technology*, July.

TEMPERLEY, C. 1948. Intermeshing Noncircular Gears. *Engineering*, **165**.

TONG, SHIH-HSI, & YANG, C. H. 1998. Generation of Identical Noncircular Pitch Curves. *ASME Journal of Mechanical Design*, **120**(June), 337–341.

WALDRON, K. J., & KINZEL, G. L. 2004. *Kinematics, Dynamics, and Design of Machinery, 2nd Edition*. John Wiley & Sons, Inc.

Index

Applications (of noncircular gears)
  crank-slider mechanism, with a, 10, 11, 13,
    174–176, 178, 180
  heavy-press machines, in, 10
  Maltese cross mechanism, with a, 7, 8
  rice planting machines, in, 6
  Scotch-Yoke mechanism, with a, 6, 11, 14, 15,
    180–183
  Tandem design with mechanisms, in, 6, 11, 13,
    15, 53, 171–178, 180–187
Axial generation, 133

Base circle, 17, 33
Bopp and Reuther's approach, 1, 9

Carrier (of planetary gear train), 138, 154
Center distance, derivation of, 26, 27, 44, 56
Centrode convexity, 30, 97, 98
Centrodes
  as closed form curves, 24, 25, 27, 118, 124
  definition of, 18
  derivation of, 21, 22, 188–200
  determination of, 18, 20, 25, 188–200
  length of, 51
Centrodes, types of
  Closed, 4
  Conventional elliptical, 43
  Eccentric circular, 94, 95
  Modified elliptical, 43, 53, 54
  Non-closed, 4, 5
  Oval, 43, 57, 60
  With lobes, 43, 60
Chordal distance, 40
Convexity, condition of, 30, 97, 98
Crank-slider mechanism, 10, 13, 171–180
Curvature of centrodes, 16, 27, 96–98
Curvature radius, 29, 36, 47
Cutting machine, 4, 5

Derivative function, 21, 22, 44, 46
Differential, 153, 154
Double enveloping process, 82

Eccentric involute gears, 6, 7, 94, 95
Eccentricity (of an ellipse), 25, 40
Ellipse parameters, 40
Elliptical gears, 40
  Conventional, 45
  External, 43
  Helical, 4, 6, 71, 92
  Modified, 11, 13, 53
  Planar, 71, 90, 91
Enveloping methods (of generation), 1, 4
Equation of meshing, 17, 71, 76–78, 80, 84–86, 88,
    90, 103, 111
Evolute, determination of, 33–36
Evolute, of tooth profile, 17, 31–33
Examples of design
  Avoidance undercutting internal gears,
    136, 137
  Generation of functions, 164, 165, 168, 169
  Helical elliptical gears, 92
  Internal noncircular gears, 118
  Planar elliptical gears, 90

Fellow's approach, 1
Final increment of function, 161
Focus (of an ellipse), 40–43
Function generation, 147, 152–155, 164, 165
Function of transmission errors, 112
Functional
  approximate solution, 163
  computational procedure, 163
Functional (of a multigear drive), 156, 161,
    162
Functional (of elliptical centrodes), 66

Generation methods
  by hob, 79, 105–112
  by master gears, 1
  by rack cutter, 1, 3, 71
  by shaper, 86, 101–104
Generation of eccentric gears, 101
  by hob, 105–112
  by shaper, 101–104

Generation of elliptical gears, 71
  by hob, 79–86
  by rack cutter, 71–78
  by shaper, 86–89
Generation of functions, 147, 152–155, 164, 165
  examples, 154, 164, 165, 168, 169, 187
  with varied sign, 153
Geneva mechanism, 7

Identity, of mating centrodes, 17
Instantaneous center of rotation, 117
Interference, 126, 132
Internal gears
  center distance, 119
  centrodes, 116–118
  derivative function, 115, 119
  generation by shaper, 126
  transmission function, 119
Internal gears (with eccentric circular pinion)
  centrodes, 123–125
  derivative function, 124
Internal gears (with oval pinion)
  center distance, 123
  centrodes, 121, 123
  transmission function, 123
Internal noncircular gears, 115
Involute, of tooth profile, 17, 31–33

Lagrange's theorem, 160, 161
Lobes, gear with, 11, 12, 60, 61, 63
Local representation, 37, 47
Localization (of bearing contact), 112
Longitudinal crowning, 112

Matrix derivation (of equation of meshing), 17,
  77, 84, 88, 103, 110
Mean Value Theorem, 160
Modified elliptical centrode (derivation), 55, 174,
  177–179
Modified elliptical gears, 11, 13, 53
Modified internal elliptical gears
  center distance, 121
  centrodes, 120
  derivative function, 120
  transmission function, 121
Monotonic (function), 152
Multigear drive, 12, 14, 147, 148, 156, 166
  assembly, 158

basic functionals, 156
centrodes, 148, 158, 159, 168
design of, 166–168
transmission function, 148

Output velocity variation, 6, 11, 53, 115
Oval gears, 9, 11, 57

Path of contact, 112
Periodic function, 118
Planetary gear train
  Elliptical gears, with, 139
    transmission function, 140
Planetary gear trains, 138–141
Polar angle, 23
Polar axis, 21, 22, 28
Polar curve, 21, 27, 28
  tangent to, 22, 23
Polar equation (of an ellipse), 41–43
Pressure angle, 37–39, 47–49
Profile crowning, 112

Rack cutter generating surfaces, 75
Rack, in mesh with, 143
  centrodes, 144
  function generation, for, 144
Relative angular velocity, 115
Rolling motion algorithm, 71–73

Satellites (of planetary gear train), 138,
  154
Scotch-Yoke mechanism, 6, 11, 14, 180–183
Shaper involute profile, 128, 129

Tandem design, 6, 11, 13, 15, 53, 147, 171–178,
  180–187
  double-crank mechanism, with, 171–174
  racks, with, 183–187
  Scotch-Yoke Mechanism, with, 180–184
  slider-crank mechanism, with, 174–179
Tooth contact analysis (TCA), 112
Tooth number, determination of, 51, 136
Transmission errors, 114
Transmission function, 27, 44, 46, 65, 66, 112

Undercutting, avoidance of, 16, 17, 47, 48,
  99–101, 126
  internal noncircular gears, of, 132, 133